Meeting America's Needs for the Scientific and Technological Challenges of the Twenty-First Century

A White House Roundtable Dialogue for President Clinton's
Initiative on Race – A Retrospective

Edited and Foreword by Oliver G. McGee, Ph.D.

Meeting America's Needs for the Scientific and Technological Challenges of the Twenty-First Century, A White House Roundtable Dialogue for President Clinton's Initiative on Race – A Retrospective

To the members of "*One America* Race Panel"

White House Office of Science and Technology Policy

and

American Association for the Advancement of Science

(1999)

Acknowledgments

"One America Race Panel"
White House Office of
Science and Technology Policy
(1999)

Dr. Howard G. Adams, Former Director of the National Institute for Mentoring at the Georgia Institute of Technology, and he is the former Director of the National Consortium for Graduate Degrees for Minorities in Engineering and Science. He is a recipient of the Presidential Award for Excellence in Science, Mathematics, and Engineering Mentoring.

Dr. John F. Alderete, Former President of the Society for the Advancement of Chicanos and Native Americans in Science (SACNAS).

Dr. Satya Atluri, Former Professor of Aerospace Engineering at University of California Irvine, Former Institute Professor of Aerospace Engineering at the Georgia Institute of Technology, former Director of the FAA Center for Aerospace Research & Education at UCLA, and member of the National Academy of Engineering.

Ms. Cathleen Barton, Former Director of Education and Work-force Strategy for the Semiconductor Industry Association.

The Late Dr. Fred Begay, Former Assistant to the President of the Navajo Government for Science & Technology.

The Late Mr. LaVelle Bond, Former Vice-President of Diversity Worldwide for Procter & Gamble.

Dr. Carlos Castillo-Chavez is a Former Professor of Biometrics at Cornell University, and is a recipient of the Presidential Award for Excellence in Science, Mathematics, and Engineering Mentoring, and a National Science Foundation Presidential Faculty Fellowship Award.

Dr. James Comer is the Maurice Falk Professor of Child Psychiatry at Yale University.

Dr. Julian M. Earls, Former Director of NASA Glenn Research Center at Lewis Field.

Dr. Darleane Hoffman, Former Charter Director of the Glenn T. Seaborg Institute for Transactinium Science at the University of California, Berkeley. Dr. Hoffman is a 1997 recipient of the National Medal of Science for her distinguished work in Nuclear Chemistry.

Dr. Shirley Malcom, Head of the Directorate for Education & Human Resources Programs for AAAS and former member of the President's Committee of Advisors on Science and Technology, and former member of the National Science Board.

The Late Dr. Samuel Massie, Former Professor Emeritus of Chemistry at the US Naval Academy. Dr. Massie held an Honoree Chair of Excellence with the U.S. Department of Energy.

Dr. Percy Pierre, Professor of Electrical Engineering at Michigan State University, Former President of Pierre View A&M, and Former Dean of Engineering at Howard University. He is a member of the National Academy of Engineering.

Dr. Richard Tapia directs the Department of Computational and Applied Mathematics at Rice University, member of the National Academy of Engineering, and former member of the National Science Board. He is a recipient of the Presidential Award for Excellence in Science, Mathematics, and Engineering Mentoring.

Dr. Lydia Villa-Komaroff is Former Vice-President of Research at Northwestern University.

The Late Dr. Herbert Z. Wong is a nationally recognized consultant of workforce diversity and inclusion pioneer for Herbert Z. Wong & Associates.

Contributors

Dr. George Campbell, Jr., Former President of Cooper Union, former President and CEO of NACME, Inc., and former member of the President's Information Technology Advisory Committee, Socioeconomic and Workforce Panel.

The Late Dr. William H. Gray, III, Former President of the United Negro College Fund and Former Majority Whip of the United States House of Representatives.

The Late Dr. Charles M. Vest, Former President and Member of the National Academy of Engineering, Former President of the Massachusetts Institute of Technology, and former member of the President's Committee of Advisors on Science and Technology.

The panel session's co-chairs, The Late Dr. David Hamburg, President Emeritus of the Carnegie Corporation of New York, and The Late Dr. John H. Gibbons, then-Assistant to the President for Science and Technology, along with Dr. Neal F. Lane, then-Director of the National Science Foundation, and later-on then-Assistant to the President for Science and Technology, succeeding Dr. Gibbons, are altogether to be commended for their efforts in pulling together a truly distinguished panel and an enormously informative conversation on the President's goal of *One America.*

Foreword

America Still Needs To Talk, Right?

Our public discourse nowadays has become so toxic around political correctness and racial insecurities that many Americans feel hesitant to reveal their real concerns about what they see is an encroachment of traditional values and meritorious exceptional core values that has characterized this nation for nearly two and half centuries. In turn we are limited in a process of silence and self-censorship.

Folks preface their dialogue on the sensitive issue of race, affirmative action, and diversity with *"my friend believes"* or *"our town is troubled about,"* in attempts to mask what is really underneath our real concerns about the state of change we really do not believe in happening across the America we once knew, but no longer is nowadays.

Underneath all of this is uncertainty and fear of what is not quite known to us - neither individually nor collectively. Few venues are available anywhere nowadays in which to have a safe real dialogue about the deepest anxieties and concerns of race, gender, and ethnicity issues facing citizens in this country.

Chat-boards online provide only temporary fixes, as reasonable (and some might say unreasonable) forums for candid discussions about race, gender and ethnicity, while at the same time, providing one with some degree of comfort in sharing ideas and issues with one another about the source of our grievances around social, technological, economic, environmental, and political (what I like to call the "STEEP") forces affecting our daily lives.

As Barbara Walters was known in her absence to "call-in" on the ladies of *The View* to counsel, we are all talking over each other instead of with other. We are so much in output mode with each other; there is no room for input from each other.

It is the sign of the times in our digital age of technological communications of instant messaging, texting, sexting, and Skyping - you name it. This is why it is so easy for online dialogues to become distorted into partial truths once inside more legitimate public forums.

Absent any contrasting even alternative knowledge centers, understandable settings, and corridors of wisdom to challenge what we are already inclined to trust, the existing prejudgments, partialities, preconceptions, predispositions, and partisanships working on credence systems of participants and institutions are only hardened.

***

Chicago law professor Cass Sunstein[1] writes about the contrasting "elements of groupthink" versus "diversity in thought" working across group dialogue comprised of sociopolitical consistency versus philosophically distinguishing points of view.

On the one hand, group dialogues of philosophically distinguishing perspective, suggests Sunstein, often are pigeonholed into discussions of anecdotal, provincial, and qualifying moderation of deep-seeding sentiments.

On the other hand, sociopolitical consistent dialogues have a tendency, Sunstein contrasts, to become more extreme in the views and perspectives exchanged, more often than not,

---

[1] Cass R. Sunstein, "The Law of Group Polarization," John M. Olin Law and Economics Working Paper No. 91 (published by the Law School, University of Chicago, 1999), http://papers.ssrn.com/paper.taf?abstract_id=199668 .

leading to a groupthink about the faultlessness, appropriateness, and even, equitableness of such views and perspectives.

Such groupthink perspectives tend to move towards the outer fringes of what we call as either "the left" or "the right" of what would be characterized as the "normal" constructive dialogue.

Sunstein's groupthink is largely seen nowadays not only in the daily forum of American politics. But also, Sunstein's groupthink is historically shattered in the form of contrarian thinking looking back in retrospective here inside this book, since President Clinton's *One America Conversations*, in forums of racial and ethnic clashes of civilizations.

The bottom-line is here is: *"Our Words Creates Our World."* We have to decide what kind of world are we creating today into tomorrow for the next-generation and the generation after that.

***

On June 13, 1997, President Clinton issued Executive Order 13050, which created the *Initiative on Race* and authorized the creation of an Advisory Board to advise the President on how to build *One America* for the 21st Century. The Board was tasked with examining race, racism, and the potential for racial reconciliation in America using a process of study, constructive dialogue, and action. The Board also focused on the role race plays in civil rights enforcement, education, poverty, employment, housing, stereotyping, courts and justice, health care, and immigration.

In his June 14, 1997 announcement launching his initiative on race headed by his seven-member distinguished multiracial, multiethnic Board (including John Hope Franklin (Chair), Linda Chavez-Thompson, Suzan D. Johnson Cook, Thomas H. Kean, Angela E. Oh, Robert Thomas, and William F. Winter), President Clinton foretold our need for real dialogue nowadays in his commencement address he delivered in late June of 1997 at the University of California at San Diego[2]. Clinton explained the purpose of *One America Conversations* was to expand upon the need for public discourse: "Over the coming year," Clinton explained, "I want to lead the American people in a great and unprecedented conversation about race."

"I want this panel to help educate Americans about the facts surrounding issues of race, to promote a dialogue in every community of the land to confront and work through these issues."

Warning of Sunstein's groupthink, Clinton went on further to presage, "Honest dialogue will not be easy at first. We'll all have to get past defensiveness and fear and political correctness and other barriers to honesty. Emotions may be rubbed raw, but we must begin."

President Clinton counseled in concluding his remarks, "But if ten years from now people can look back and see that this year of honest dialogue and concerted action helped to lift the heavy burden of race from our children's future, we will have given a precious gift to America."

The *One America* Advisory Board's Report to the President, entitled "*One America in the 21st Century – Forging a New Future*," was submitted to the White House and the public in September 1998. During the prior 15 months, board members brought together leaders from religious groups and corporate sectors for visioning sessions on "seeking ways to build a more united and just America," and for listening forums with Ameri-

---

[2] President William Jefferson Clinton, Commencement Speech, University of California at San Diego, June 14, 1997, http//www.whitehouse.gov/Initiatives/One America/announcement.html.

cans "who revealed how race and racism have impacted their lives."

The Board's efforts were not intended as a definitive assessment of the state of race relations in Clinton's America. For his *One America* Board was given no independent authority to commit Federal resources to address race in America. Rather, the *One America* Board gave anecdotal impressions of its national outreach – including searches for common ground, struggles with the legacy of race and color, bridging the gaps for capacity building in race relations and civil rights across American institutions of family, schools, churches, government, corporations, and philanthropy, and finally, development of long-term strategies for the future to advance America's race relations in the 21st century – for any subsequent White House implementation and/or any potential congressional action.

Looking back in retrospective today, *One America Conversations* received mixed reviews, as a whole, on its overall impact on racial and ethnic clashes of civilizations across America, albeit it did usher in the Age of Obama, whose reviews are equally mixed as recent polls reveal at the time of this writing. The distinguished black historian The Late John Hope Franklin would be proud of the resulting outcome of his panel leadership culminating in the Age of Obama more than ten years hence. However, the tone launched by his *One America* deliberations was restrained by the panel's hearing of testimony only of people who supported the value of diversity.

Since the historic *One America Conversations*, challenges facing racial and ethnic clashes of civilizations in American society remain today.[3]

---

[3] Oliver McGee (ed.), *"Future of African American Men and Boys, Promoting the Saving, Transforming and Empowering of African American Men and Boys for the Betterment of American Society,* A Roundtable Dialogue of a *Commission on Strategizing the Future of African American Men and Boys,* (©2011 Oliver G. McGee III, Revilo Group, L.L.C.)

FUTURE OF AFRICAN AMERICAN MEN & BOYS

Promoting the Saving, Transforming and Empowering
of African American Men and Boys
for the Betterment of American Society

Edited by Oliver G. McGee, Ph.D..

***

According to a recent 2007 Roundtable Dialogue of a *Commission on Strategizing the Future of African American Men and Boys*, African American men and boys face many challenges in the United States.

In response to these challenges, Morgan State University and The Kellogg Foundation co-sponsored a dialogue on June 26-27, 2007 at the Lansdowne Resort Boardroom in Lansdowne, Virginia entitled, *Future of African American Men and Boys: Promoting the Saving, Transforming and Empowering of African American Men and Boys for the Betterment of American Society.*

The Lansdowne dialogue was co-chaired by Dr. Calvin O. Butts III, President, SUNY Old Westbury, and Dr. Earl Richardson, then-President, Morgan State University. Butts and Richardson also served as co-moderators of the conversation.

This Lansdowne dialogue brought together 12 prominent African American leaders to inform business, government, university, and philanthropic interests on suitable goals and strategies for improving and enhancing the lives of African American men and boys in American society.

Historically, there has been a significant gap between African American males and the rest of society. Recent empirical studies suggest that closing the gap in how black male youngsters learn in particular has solutions, as described in the **Sidebar: Study Finds That Black Youngsters Perform Better Academically in Different Types of Learning Environments Than Do Whites**, Source: *The Journal of Blacks in Higher Education*, June 18, 2009.

The statistics are staggering and represent a compelling reason for these matters to be addressed on local, regional and national levels. As noted in Raymond Winbush's book[4]:

§   The life expectancy of African American males in Washington, D.C. is 57.3 years. Only seven other nations in the world have a lower life expectancy for their male population: Bangladesh (53.5), Ethiopia (51.5), Myanmar (54.5), Pakistan (56.5), Sudan (53.0), Tanzania (52.5), and Zaire (54.0).

§   A black male has a 1 in 20 chance of being imprisoned while in his twenties.

§   A black male has a 1 in 2 chance of not attending college even if he graduates from high school.

§   A black male has a 1 in 3,700 chance of getting a Ph.D. in mathematics, engineering, or the natural sciences.

§   A black male has a 1 in 766 chance of becoming an attorney.

§   A black male has a 1 in 395 chance of becoming a physician.

Other devastating statistics[5] regarding the plight of the black male include the following:

---

[4] Raymond Winbush, *The Warrior Method: A Program for Rearing Healthy Black Boys* (Amistad Press, 2001)

[5] Oliver McGee (ed.), "*Future of African American Men and Boys, Promoting the Saving, Transforming and Empowering of African American Men and Boys for the Betterment of American Society,* A Roundtable Dialogue of a

- Only 48% percent of African American households are married-couple families compared to 77% for the general population. Forty-four percent (44%) of African American households are maintained by women with no spouse present so many African American males are growing up in single-parent households with no positive male role model.

- Thirty-two percent (32%) of African American males age 18 years and younger live in poverty compared to 17% of all males in this age group.

- Eight percent (8%) of African American males, age 18 years or less and 11% of African American males age 18-29 years are in supervised care or custody (correctional or juvenile institutions) compared to 1% and 3% respectively for the general male population in these age groups.

- Only 16% of African American males age 25 years and over had at least a college degree compared to 28% of all males in this group [see **Sidebar: Black Progress in Winning Professional Degree Awards**, Source: *The Journal of Blacks in Higher Education* (2009)].

- Educational attainment impacts labor force roles as only 18% of African American males age 16 years and over occupy managerial and professional positions compared to 29% of all males; African American males are more prominent in service and unskilled labor occupations.

−   In 2001, African American children (17 years old and younger) were twice as likely to lack health insurance coverage versus white, non-Hispanic children: 13.9% versus 7.4% respectively with lack of health insurance more common among the poor (30.7%).

−   In 2001, 38.3% of African American children were covered under Medicaid insurance versus 15.3% of non-Hispanic white children.

−   In 2000, 15.3% of African American males 16-24 were high school drop outs versus 7.0% of white males.

−   In 1999, African American students lagged behind white, non-Hispanic for mathematics proficiency levels:

- 9-year olds: 63.3% of African Americans versus 88.6% of whites had beginning skills and understanding.
- 13-year olds: 50.8% of African Americans versus 86.7% of whites were proficient with numerical operations and beginning problem solving.
- 17-year olds: 26.6% of African Americans versus 69.9% were proficient with moderately complex procedures and reasoning.

−   In 1999, African American students lagged behind white, non-Hispanics for reading proficiency levels:

- 9-year olds: 36.0% of African Americans versus 73.0% of whites were able to understand, combine ideas, and make inferences based on short uncomplicated messages.
- 13-year olds: 38.0% of African Americans versus 69.0% of whites were able to search for specific information, inter-relate ideas and make generalizations.
- 17-year olds: 66.0% of African Americans versus 87.0% of whites were able to search for specific information, interrelate ideas and make generalizations while 17.0% of African Americans and 46.0% of whites had advanced reading proficiency skills.

These problems have attracted well-intended, if sporadic, attention in the philanthropic community and occasionally have prompted government or business focused initiatives. From time to time there have been efforts made to address underlying causes behind this complex and multi-dimensional issue, but such programs are often limited to bridging gaps or treating disparities in one sector of society (such as education or health care), while others are ignored. They do not offer a comprehensive, multi-sector and systemic approach.

Furthermore, the language that surrounds a discussion of the Commission on African American Men and Boys and their roles suggests strongly that "access and opportunity" constitute a full remedy for the lack of engagement that plagues American society. Opportunity is crucial and it is often lacking, [see **Sidebar: Only One of the 278 Winners of the 2009 Goldwater Scholarships Is Black**, Source: *The Journal of Blacks in Higher Education* (2009)], including gaps that persist in the achievement of higher educational entrance requirements to such opportunity [see **Sidebar: "This Was Not Supposed to Happen: The Black-White Test Gap Is Growing,"** Source: *The Journal of*

Blacks in Higher Education 25 (1999)], but the challenge is one of engagement and leadership [see **Sidebar: Norfolk State University Seeks Out Black Men**, Source: *The Journal of Blacks in Higher Education* (2009)], not simply opening gates where there is no road.

***

Finally, the national and international political environments for these issues are dramatically changing. Given the demographic certainties of the next three decades, forfeiture of the leadership, creativity, industry and engagement of African American men and boys in U.S. society will be disastrous economically, politically and culturally as well as to our families, communities and institutions. The continued alienation of a major segment of our society constitutes a very clear and immediate danger to our ability to survive in freedom. No democracy can withstand the ongoing systemic cultivation of injustice, or manage deepening resentment within its populations without resorting to divisive fear and social controls.

This last point bears elaboration. The current "civil response" to our long-term national failure to engage young men of color is too often incarceration and isolation. The consequences of this policy (and it must be recognized as such) are well understood. This policy is not sustainable for many reasons. It is destructive and wasteful at all levels of our economic and community lives. But this systemic failure is also inherently dangerous in today's international context. We are sowing seeds of anger and alienation at a time when our national security depends on unity and trust.

Nonetheless, scientists and engineers continue to have a unique somewhat neutral contribution to make to the challenges identified by President Clinton's *Initiative on Race*, particularly in regards to the development of talent among groups traditionally underrepresented in science, mathematics, and engineering.

In response to these challenges back in 1997, the Office of Science and Technology Policy (OSTP) and the Directorate for Education and Human Resources Programs of the American Association for the Advancement of Science (AAAS) co-sponsored a *One America* panel session at the AAAS 150th Anniversary Meeting held February 13, 1998 entitled, *Meeting America's Needs for the Scientific and Technological Challenges of the Twenty-First Century.* The OSTP-AAAS *One America* panel was co-chaired by The Honorable John H. Gibbons, then-Assistant to the President for Science and Technology, and David Hamburg, then-President Emeritus of the Carnegie Corporation of New York and former member of the President's Committee of Advisors on Science and Technology (PCAST). Hamburg also served as the moderator of the conversation. The panel session brought together twenty prominent scientists and technological leaders from academia, industry, laboratories, professional societies, and government to discuss the need for participation by all Americans in science and technology, and the responsibilities and expectations of a diverse scientific and technological community in contributing to the national good through research and education.

In retrospective today, looking back over two decades hence, some advancements in diversity enhancement and work-force development of American enterprise has been achieved with lukewarm mixed results, requiring government, universities and industry to create substantially more advancements and innovations, since the 1999 White House Office of Science and Technology Policy's *One America Conversation* on *Meeting America's Needs for the Scientific and Technological Challenges of the 21st Century.*

As President Clinton stated then, "First, science and its benefits must be delivered toward making life better for all Americans – never just a privileged few ... Science must not create a new line of separation between the haves and the have-nots, those with and those without the tools and understanding to learn and use technology ... Science can serve the values and interests of all Americans, but only if all Americans are given a chance to participate in science."

A diverse scientific and technological workforce ultimately positions the United States for continued leadership in this century. Historically speaking, tenets of this panel conversation rested on Executive Order 11246 by President Johnson on September 24, 1965, "establishing requirements for non-discriminatory practices in hiring and employment on the part of the United States government contractors." In other words, President Johnson's Executive Order mandating affirmative action compliance in government-industry-university partnerships within specific timetables and timelines in order to do business with the federal government including university research capacity building. This put the teeth in President Kennedy's Civil Rights Act of 1964 signed by President Johnson in Kennedy's legacy. Upon sun down of Executive Order 11246 three decades later, the Clinton Administration famously pronounced, "let's mend it don't end it" in its discussions with the nation about the future of affirmative action. This brought about the core essence of the *One America Conversations*.

The richness and foresight of this historical panel's input is rooted in the experience and diversity of the truly distinguished participants. The enormously informative conversation on President Clinton's goal of *One America* helped inform the Clinton Administration, and perhaps serves to inform the current and future administrations on suitable goals and strategies for diversifying the scientific and technological community. As we leave the age of President Obama and engage the age of President Trump, this retrospective takes a sober look at how far we have come, the substantial lack progress we have achieved, and the new issues and challenges that now frame the social, technological, education, economic and political (STEEP) conversations on *"Getting to 2076 – America's Tercentennial,"* which is happening now across America's social fabric in the digital age of demography shift and heightened engagement of our citizenry.

***

Sidebar: "Study Finds That Black Youngsters Perform Better Academically in Different Types of Learning Environments Than Do Whites"

Source: *The Journal of Blacks in Higher Education* (2009)

"A recent study published in *Cognition and Instruction* finds that black students in fourth and fifth grade perform better academically in certain types of learning environments. The authors of the study are A. Wade Boykin, a professor of psychology at Howard University, Brenda A. Allen, recently named provost and professor of psychology at Winston-Salem State University, and Eric A. Hurley, an assistant professor of psychology and black studies at Pomona College.

Researchers divided a large group of fourth- and fifth-graders at an urban school in the Northeast and placed them in three different learning environments. One group was placed in a communal learning environment where they were urged to work together to solve problems. A second group was told they would earn an award if the combined performance of the group exceeded expectations. The third group was told that those individuals who performed the best would be rewarded.

The results showed that black students performed best in the communal group. The black students did the worst in the third group that emphasized individual achievement. White students, on the other hand, performed the best in the group that emphasized individual competition and did the worst in the communal group.

Professor Boykin believes that black students perform better in the communal group setting because of the tendency in the black community to have large extended families and more involvement in community-based institutions, which make them more comfortable in a group-learning environment.

The study raises the question of whether the racial gap in standardized test scores and other measures of academic measure can be narrowed by changing the way black children are taught in the public schools."

***

Sidebar: "Only One of the 278 Winners of the 2009 Goldwater Scholarships Is Black"

Source: *The Journal of Blacks in Higher Education* (2009)

"Barry Goldwater, the famed conservative U.S. senator from Arizona and 1964 GOP presidential nominee, was a dedicated promoter of scientific and engineering research. In 1986, when Congress created a new scholarship program to encourage graduate study in mathematics, science, and engineering, Goldwater's name was attached to the new program.

Students chosen as Goldwater Scholars can obtain tuition grants of $7,500 per year for two years. Since its founding the program has awarded more than 5,800 scholarships with a total value of $56 million.

Very few blacks have benefited from the Goldwater Scholarship program. But racism is not the culprit. The low number of black students pursuing graduate study in the sciences who meet the eligibility requirements results in a small pool of black applicants.

This spring 278 Goldwater Scholars were selected from a pool of 1,097 applicants. The Goldwater Scholarship Foundation announced that only one of the 278 Goldwater Scholars self-identified as an African American.

Gerald Smith, the president of the Goldwater Scholarship Foundation, announced that black students with the academic qualifications and graduate school aspirations required for Goldwater Scholarships are quickly "snapped up" by the nation's leading research universities. Often they receive full-tuition scholarships and have no need to seek additional cash awards to finance their higher education."

***

Sidebar: "Black Progress in Winning Professional Degree Awards"

Source: *The Journal of Blacks in Higher Education* (2009)

"In the 2006-07 academic year blacks earned 6,474 professional degrees. These made up 7.2 percent of all professional degrees awarded in the United States that year. These include degrees in medicine, law, dentistry, and several other fields.

More than 3,100 African Americans earned a law degree in the 2006-07 academic year, making up 7.3 percent of all law degree recipients. They were nearly half of all blacks who earned a professional degree. More than 1,100 black students earned a medical degree, making up 7.2 percent of all medical school graduates. Blacks made up more than 17 percent of all students who earned a professional degree in podiatry and nearly 14 percent of all students who won a professional degree in divinity.

However, blacks continue to have a very small presence in professional degree awards in dentistry, osteopathic medicine, optometry, chiropractic medicine, and veterinary medicine."

***

Sidebar: "Norfolk State University Seeks Out Black Men"
> Source: *The Journal of Blacks in Higher Education* (2009)

"Increasingly black women are becoming a larger percentage of the student bodies at historically black colleges and universities. For example, women now make up 73 percent of all undergraduate enrollments at Clark Atlanta University.

At Norfolk State University in Virginia, women make up 62 percent of the undergraduate student body. But Norfolk State is determined to boost black male enrollments. The university is now offering $2,500 annual scholarships to black men who graduate from local high schools with a 3.0 grade point average. Black women are not eligible for the scholarships. The scholarships resulted in an increase of 27 percent in black male freshmen this past academic year."

***

Sidebar: Historically Black Colleges and Universities (HBCU) Week – September 14-18, 2010

> Source: *The Huffington Post* (2010)
> Source: Posted September 14, 2010, by April Ryan, White House correspondent, *American Urban Radio Networks*

"President Obama has proclaimed this HBCU Week. The 100 plus Historically Black Colleges and Universities are a single source for propelling blacks into the middle class.

Student funding and retention are key for HBCU's in the bad economy. Dr. Michael Lomax of the United Negro College Fund says over 50 percent of students drop out of college with many citing financial difficulty.

Dr. Robert Franklin, President of Morehouse College says, "The national graduation rate for African American men is lower than any other segment of the population." He contends "finances keep many of our students away from the finish line." He says as little as six thousand dollars per student makes the difference. The Atlanta School, which boasts Dr. Martin Luther King Jr. as a graduate, has an annual price tag of 35 thousand dollars.

Dr. Lomax contends, "the average cost of a private HBCU is 20 thousand dollars a year, consistently 30 percent lower than an equivalent white institution."

In a related note, the United Negro College Fund says it motto "*A Mind is a Terrible Thing to Waste*" turns 40 next year."

***

Sidebar: "This Was Not Supposed to Happen: The Black-White Test Gap Is Growing"

Source: *The Journal of Blacks in Higher Education* 25 (1999)

"A large gap persists in the ensuring decade after this important article between white and black standardized test scores remarkably still nowadays for the following reasons:

"[B]lack students who take the SAT have not followed the same academic track as white students … white SAT test takers are more likely than black SAT takers to have completed courses in geometry. In higher level mathematics, such as trigonometry and calculus, whites hold a large lead. In 1999, 52 percent of white SAT takes had taken trigonometry in high school compared to 40 percent of black test takers. A full one-quarter of white test takers had taken calculus in high school. Only 13 percent, about half as many, of black students had taken calculus."

Analogous gaps can be seen in the preparation for the verbal portion of standardized tests, where black students fall short in literature studies preparation and advanced writing skills. In addition, there are discrepancies of wide significance of fewer black students enrolling in test preparation programs from *Kaplan* and the *Princeton Review*, which purport at least a 100 point increase in overall test scores upon completion and are highly recommended by the *Education Testing Service* (ETS). Diversity advocacy groups and foundations like the Gate Foundation contribute billions of dollars through its *Gates Millennium Scholarship* in funding standardized test preparation programs for financially disadvantaged black students [Theodore Cross, "Bill Gates' Gift to Racial Preferences in Higher Education," *Journal of Blacks in Higher Education* 25, 6-7 (1999)]."

***

Sidebar: "Work is Plentiful for Black Engineers – Top candidates sought despite slow economy"

Source: *The Columbus Dispatch*, May 1, 2011
Source: by Steve Giegerich, *St. Louis Post-Dispatch*

"ST. LOUIS – Brenda Nathan doesn't shy away from the perception of her chosen profession – she embraces it.

"I'm a complete nerd," boasted the mechanical engineering major from California Polytechnic State University. "I take pride in it."

Nathan was one of 8,000 young people attend[ed] the National Society of Black Engineers convention in St. Louis recently.

The revenge of the nerds – as Nathan and other conventioneers can attest after four days of courtship by the country's top corporations – is called job opportunity.

In addition to the job fair, the four-day event featured symposiums on employment ("The Parallels of Job-Seeking and Dating"), personal enhancement ("Eliminating Self-Defeating Behavior") and complex engineering discourse ("Battle of the Frameworks: ITIL/Six-Sigma/ISO 9000/CMMI").

Organizers said the presence of 300 top-drawer exhibitors – Apple, Boeing, Intel, the CIA, Johnson & Johnson, Honda, Facebook and General Mills to name several – represented more than an empty gesture to minority hiring. Several companies, in fact, reserved interview rooms off the convention floor with the express purpose of making on-the-spot offers to qualified candidates.

Taylor Armstead, a fourth-year electrical-engineering major from Hampton University in Virginia, emerged from a get-acquainted conversation with Intel recruiters. He was hoping to be among those departing St. Louis with a job in hand.

Though the African American unemployment rate nationwide continues to run far higher than that for the general population – 15.7 percent compared with 8.9 percent – highly qualified minority candidates remain in demand in white-collar professions conscious of diversity.

Armstead is fully aware of the advantage of being a highly educated black candidate: "We're a hot commodity right now," he said.

Nationally, 2 million people earn a living as professional engineers. Of those, the engineers group estimates 70,000 are African American.

The mission of the engineers group is to encourage parents and secondary and post-secondary institutions to better prepare young people for the rigors of a demanding field of study based on science and math. Success in an engineering curriculum is predicated on passing three key courses - calculus, chemistry and physics, said Carl Mack, executive director of the 35,000-member society of black engineers.

Seventy percent of the 10,000 black students who enroll as engineering majors each fall don't make the cut, he said.

Mack said it is incumbent on blacks, especially schools, to turn young people on to math and science as a practical alternative to the minuscule odds of striking it big in sports or entertainment.

To Mack, the group's outreach effort is the not-so-simple task of getting enough kids, and their parents, to buy into the reality of education being its own reward.

A relieved Ashley Jackson is now able to boast that she survived the gantlet of calculus/chemistry/physics. "I won't say it was the easiest thing, but I made it," said the biomedical-engineering major from Virginia Commonwealth University.

Jackson said she doesn't see a lot of classmates that look like her in VCU engineering-school classrooms.

"I can probably count them on one hand," she said.

But the transformation is under way in a profession once almost exclusively the domain of white males.

"Things are changing," said Charmaine Flemming, a graduate student in chemical engineering at the Florida Institute of Technology.

The proof of that is the engineering group itself. Women now make up more than 40 percent of the membership."

***

Sidebar: Here's What The National Science Board Says About The Future of America's Workforce Development in Science, Technology, Engineering and Mathematics

THE SKILLED TECHNICAL WORKFORCE:

Crafting America's Science & Engineering Enterprise

"Many know NSF's mission is to advance the progress of all fields of science and engineering. They may be less familiar with the agency's mandate to prepare the STEM workforce of the future, including the skilled technical workforce. In addition to our Advanced Technical Education program, which educates highly qualified S&E technicians for advanced technology fields through community college support, NSF supports the STW in numerous other ways, such as through education research and our Convergence Accelerator on the Future of Work."

France Córdova
NSF Director

"Businesses large and small across the U.S. need adaptable, STEM-capable workers at every education level and from all demographic groups in order to be competitive. Creating a strong, diverse, STEM-ready workforce is essential to economic and social prosperity and we all have a role to play in this critical effort."

-Victor McCrary

NSB member,
VP for Research and Economic Development
at Morgan State University.

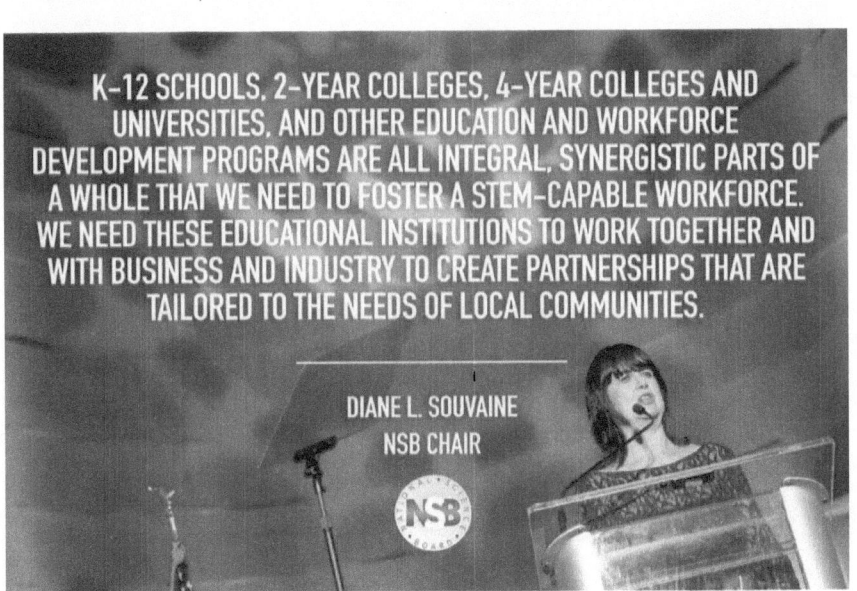

K-12 SCHOOLS, 2-YEAR COLLEGES, 4-YEAR COLLEGES AND UNIVERSITIES, AND OTHER EDUCATION AND WORKFORCE DEVELOPMENT PROGRAMS ARE ALL INTEGRAL, SYNERGISTIC PARTS OF A WHOLE THAT WE NEED TO FOSTER A STEM-CAPABLE WORKFORCE. WE NEED THESE EDUCATIONAL INSTITUTIONS TO WORK TOGETHER AND WITH BUSINESS AND INDUSTRY TO CREATE PARTNERSHIPS THAT ARE TAILORED TO THE NEEDS OF LOCAL COMMUNITIES.

DIANE L. SOUVAINE
NSB CHAIR

Individuals employed in S&E occupations in the United States: Selected years, 1960–2017

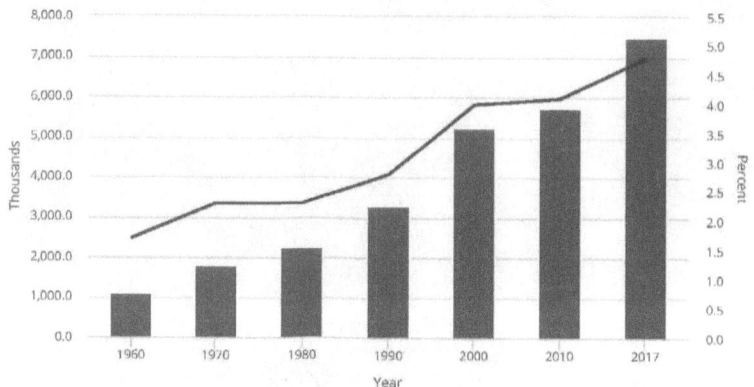

● Number (thousands) — S&E as a percent of all U.S. occupations

Note(s)
Data include individuals at all education levels.

Source(s)
Census Bureau, Decennial Census, 1960–2000; Minnesota Population Center, Integrated Public Use Microdata Series, International: Version 7.1, Minneapolis, MN: IPUMS (2018); and American Community Survey (ACS), 2010, 2017, Public Use Microdata Sample (PUMS).

Science and Engineering Indicators

Estimated number of researchers in selected regions, countries, or economies: 2009–16

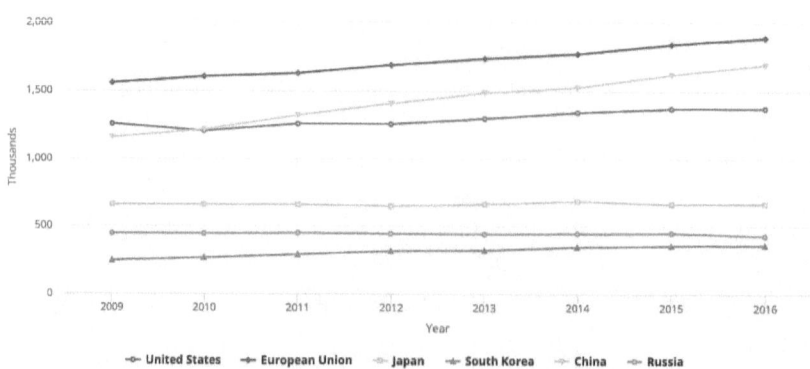

-●- United States -●- European Union -○- Japan -▲- South Korea -○- China -○- Russia

Note(s)
Researchers are full-time equivalents.

Source(s)
Organisation for Economic Co-operation and Development, Main Science and Technology Indicators, 2018/2 (2019).

Science and Engineering Indicators

Gross Expenditures on R&D

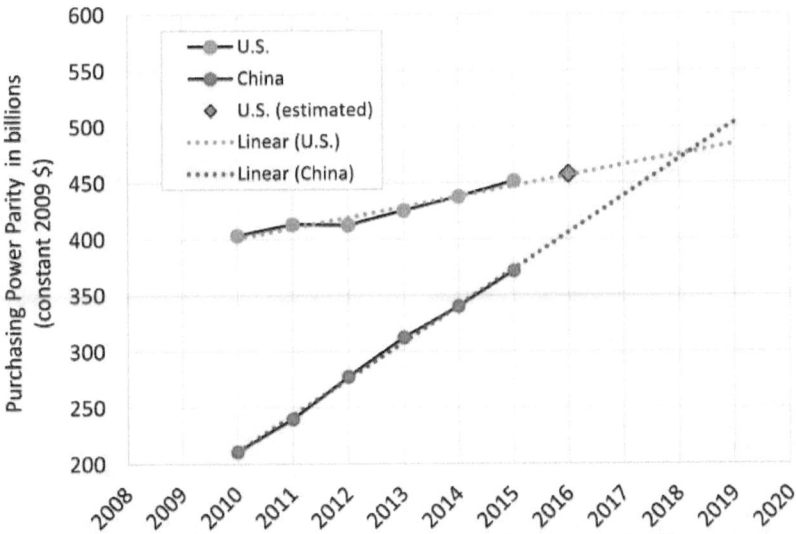

SOURCES: All data are from *Science and Engineering Indicators 2018* except for the U.S. estimated R&D expenditures for 2016, which are from the National Center for Science and Engineering Statistics *National Patterns of R&D Resources, 2017.*

The Skilled Technical Workforce
By Race and Ethnicity: 2017

- Overall Workforce ■ Skilled Technical Workforce

White non-Hispanic	64%	66%
Hispanic	17%	18%
Black	11%	10%
Asian	6%	4%
Other	2%	2%

The Skilled Technical Workforce
By Sex: 2017

- Overall Workforce ■ Skilled Technical Workforce

Male	47%	72%
Female	53%	28%

National Science Board, "Science and Engineering Labor Force," *Science and Engineering Indicators 2020* (forthcoming).
Data source: Census Bureau, American Community Survey 2017, public use microdata

The Skilled Technical Workforce
By Sex: 2017

■ Overall Workforce ■ Skilled Technical Workforce

Male	47% 72%
Female	53% 28%

National Science Board, "Science and Engineering Labor Force," *Science and Engineering Indicators 2020* (forthcoming).
Data source: Census Bureau, American Community Survey 2017, public use microdata

Women in the workforce and in S&E: 1993 and 2017

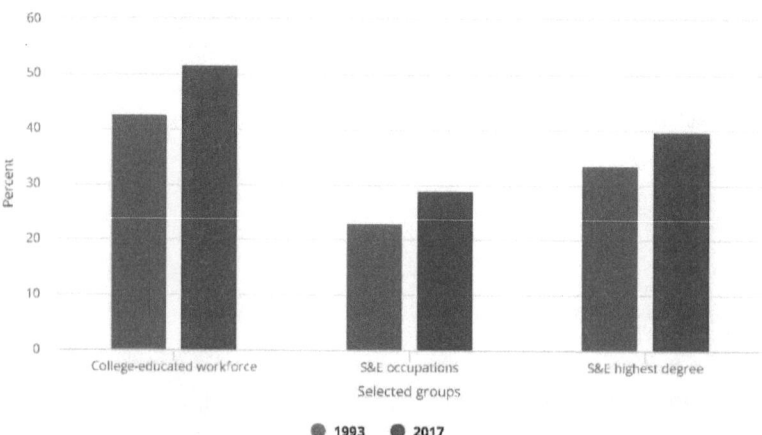

● 1993 ● 2017

Source(s)
National Center for Science and Engineering Statistics, National Science Foundation, Scientists and Engineers Statistical Data System (SESTAT), 1993,
and the National Survey of College Graduates (NSCG), 2017.

Science and Engineering Indicators

Representation of racial and ethnic groups in the U.S. population and among S&E degree recipients: 2017

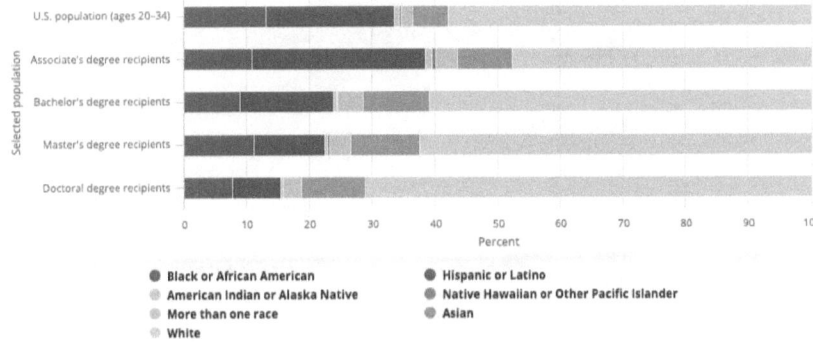

Note(s)
Hispanic may be any race; race categories exclude Hispanic origin. U.S. population data reflect the percentage of people in each racial and ethnic group in the U.S. population between ages 20 and 34 on 1 July 2017. Degree totals may differ from those elsewhere in the report; degrees awarded to people of unknown or other race were excluded.

Source(s)
U.S. population data from the U.S. Census Bureau. Degree data from National Center for Education Statistics, Integrated Postsecondary Education Data System (IPEDS), Completions Survey.

Science & Engineering Indicators

Broad S&E occupational categories, by employment sector: 2017

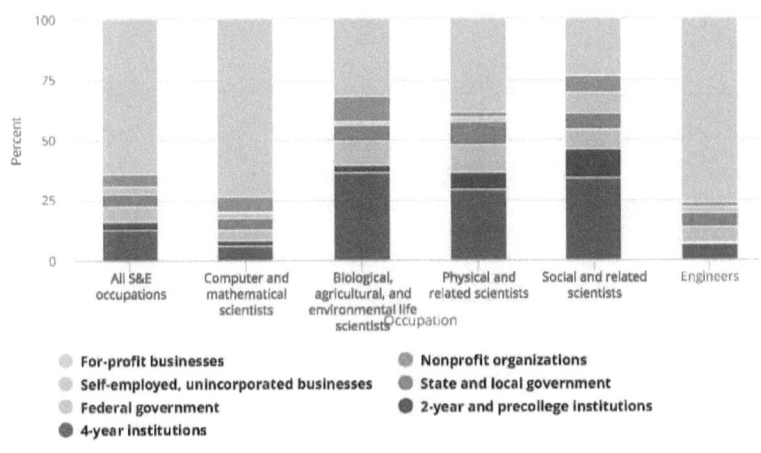

Note(s)
Percentages may not add to 100% because of rounding.

Source(s)
National Center for Science and Engineering Statistics, National Science Foundation, National Survey of College Graduates (NSCG), 2017.

Science and Engineering Indicators

Estimated salary differences between women and men with highest degree in S&E employed full time, controlling for selected characteristics, by degree level: 2017

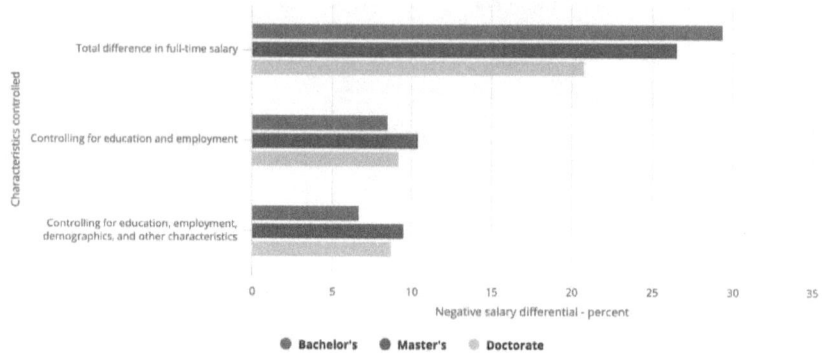

Note(s)

Salary differences represent the estimated percentage difference in women's average full-time salary relative to men's average full-time salary. Coefficients are estimated in an ordinary least squares regression model using the natural log of full-time annual salary as the dependent variable, then transformed into percentage difference. Controlling for education and employment includes 20 field-of-degree categories (out of 21 S&E fields), 38 occupational categories (out of 39 categories), 6 employment sector categories (out of 7 categories), years since highest degree, and years since highest degree squared. In addition to the above education- and employment-related variables, controlling for education, employment, demographics, and other characteristics includes the following indicators: nativity and citizenship, race and ethnic minority, marital status, disability, number of children living in the household, geographic region (classified into 9 U.S. Census divisions), and whether either parent holds a bachelor's or higher-level degree.

Source(s)

National Center for Science and Engineering Statistics, National Science Foundation, National Survey of College Graduates (NSCG), 2017, and the Survey of Doctorate Recipients (SDR), 2017.

Science and Engineering Indicators

Foreign-born scientists and engineers employed in S&E occupations, by highest degree level and broad S&E occupational category: 2017

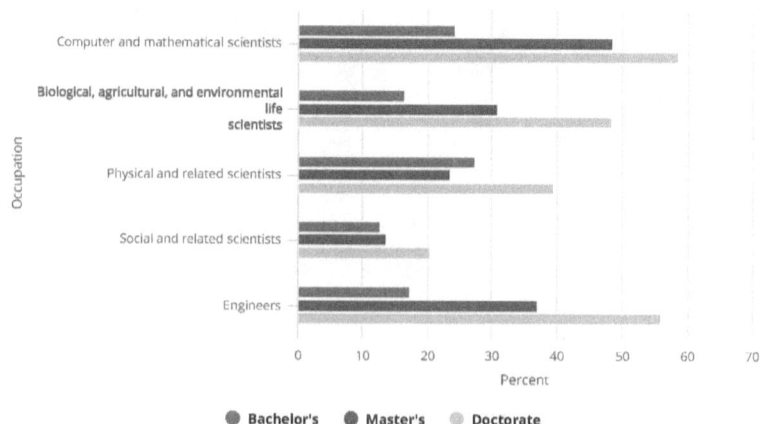

Source(s)

National Center for Science and Engineering Statistics, National Science Foundation, National Survey of College Graduates (NSCG), 2017.

Science and Engineering Indicators

Estimated number of researchers in selected regions, countries, or economies: 2009–16

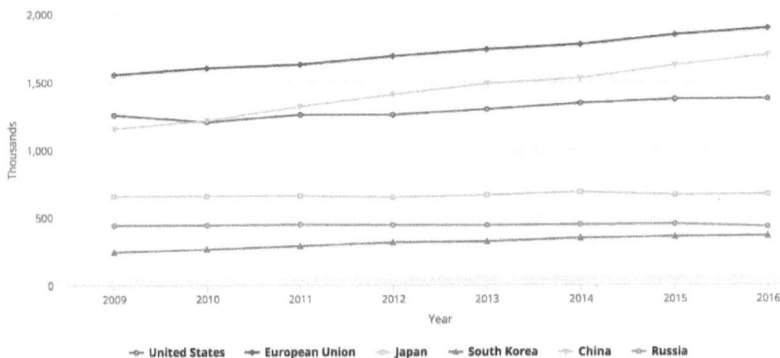

Note(s)
Researchers are full-time equivalents.

Source(s)
Organisation for Economic Co-operation and Development, Main Science and Technology Indicators, 2018/2 (2019).

Science and Engineering Indicators

Immediate college enrollment rates among high school graduates, by institution type: 1980–2016

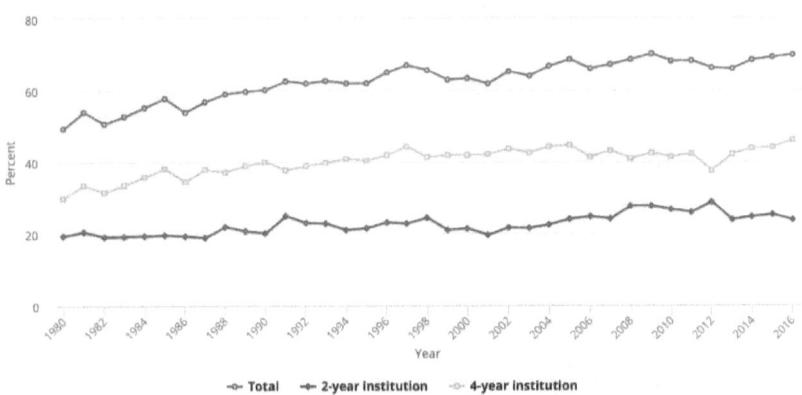

Note(s)
The figure includes students ages 16–24 who completed high school in each survey year. Immediate college enrollment rates are defined as rates of high school graduates enrolled in college in October after completing high school. Before 1992, high school graduates referred to those who had completed 12 years of schooling. As of 1992, high school graduates are those who have received a high school diploma or equivalency certificate. Detail may not add to total due to rounding.

Source(s)
McFarland J, Hussar B, Wang X, Zhang J, Wang K, Rathbun A, Barmer A, Forrest Cataldi E, and Bullock Mann F, *The Condition of Education 2018*, NCES 2018-144 (2018), Tables 302.10, 302.20, 302.30. See Table S1-6.

Science and Engineering Indicators

Employed underrepresented minorities with highest degree in S&E, by degree level: 1993–2017

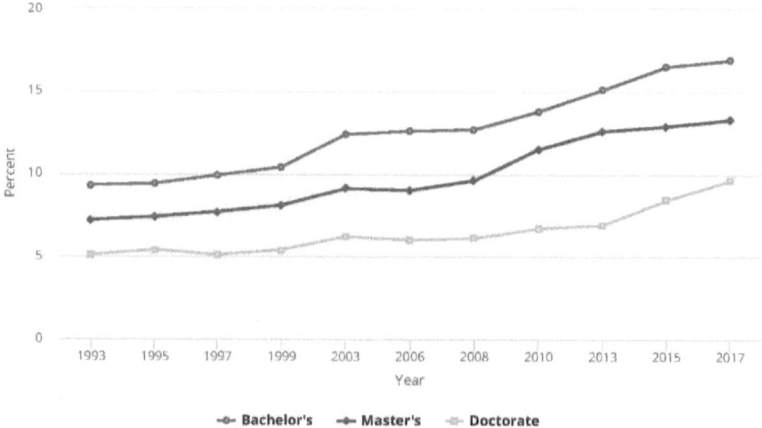

Bachelor's · Master's · Doctorate

Note(s)
Underrepresented minorities include blacks or African Americans, Hispanics or Latinos, and American Indians or Alaska Natives. Hispanic may be any race; race categories exclude Hispanic origin.

Source(s)
National Center for Science and Engineering Statistics, National Science Foundation, Scientists and Engineers Statistical Data System (SESTAT), 1993–2013, and the National Survey of College Graduates (NSCG), 2015–17.

Science and Engineering Indicators

Women in S&E occupations: 1993–2017

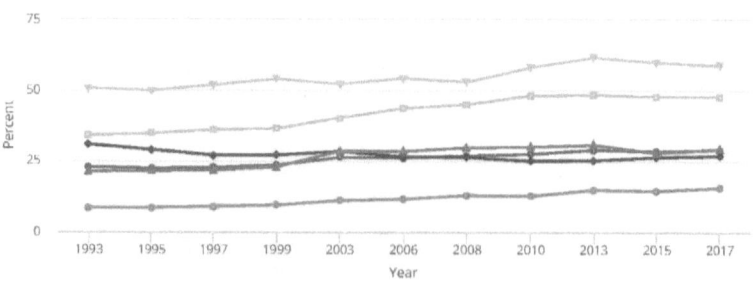

· All S&E occupations
· Computer and mathematical scientists
· Biological, agricultural, and environmental life scientists
· Physical scientists
· Social scientists
· Engineers

Note(s)
National estimates were not available from the Scientists and Engineers Statistical Data System (SESTAT) in 2001.

Source(s)
National Center for Science and Engineering Statistics, National Science Foundation, SESTAT, 1993–2013, and the National Survey of College Graduates (NSCG), 2015–17.

Science and Engineering Indicators

FIGURE 3-25

Foreign-born individuals with highest degree in S&E living in the United States, by place of birth: 2017

S&E highest-degree holders

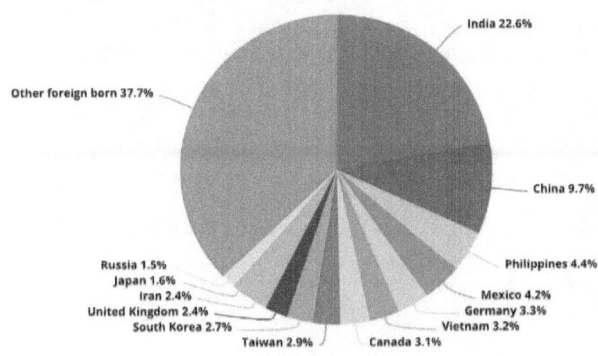

S&E doctorate holders

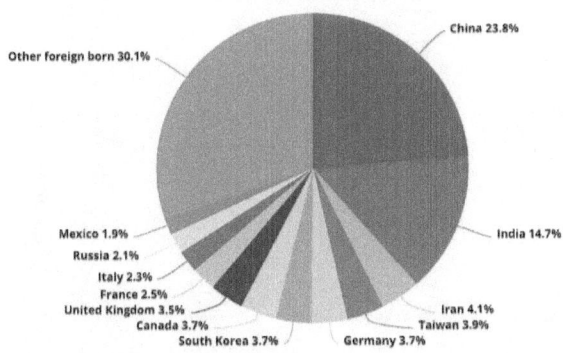

Source(s)
National Center for Science and Engineering Statistics, National Science Foundation, National Survey of College Graduates (NSCG), 2017.

Science and Engineering Indicators

THE WORKFORCE THAT ACTUALLY BUILT THE SPACE SHUTTLE WERE NOT FOUR-YEAR DEGREE ENGINEERS, THEY WERE SKILLED TECHNICIANS. WE NEED TO MAKE SURE THOSE JOBS ARE REPRESENTED.

– DR. MAE JEMISON, PRINCIPAL, 100 YEAR STARSHIP; NASA ASTRONAUT (RETIRED)

***

Have Diversity Initiatives Made Us More Divisive?

Chief Diversity Officers inside modern enterprises first took off at the start of the millennium. Institutions across the social fabric (including governments, corporations, universities, and charities) viewed these officers as ones that would: (1) impact their organizations in a positive way, including enterprise value for shareholders and stakeholders, (2) raise sensitivity and awareness around diversity and inclusion, and (3) create a workplace that valued diversity, which could then permeate across the social fabric.

Photo Credit: Liberty News

According to Scientific American on September 16, 2014, "The first thing to acknowledge about diversity is that it can be difficult. In the U.S., where the dialogue of inclusion is relatively advanced, even the mention of the word "diversity" can lead to anxiety and conflict. Supreme Court justices disagree on the virtues of diversity and the means for achieving it. Corporations spend billions of dollars to attract and manage diversity both internally and externally, yet they still face discrimination lawsuits, and the leadership ranks of the business world remain predominantly white and male."

Photo Credit: Lyndon B. Johnson speaks to the nation on July 2, 1964 before signing the Civil Rights Act of 1964. East Room, The White House, Washington, DC.

This article more specifically addresses exactly what this premise raises squarely and frankly to engender water-cooler discussion and further comment by you. For diversity is more than just skin deep, it is about people first and foremost. Has diversity divided us into divisiveness is the issue raised herein. What have these diversity and inclusion programs accomplished that truly permeates across the social fabric?

***

Diversity and the Social Fabric

May 17, 2018 marks the 64th anniversary of the 1954 *Brown vs. Board of Education* landmark Supreme Court ruling outlawing segregation in public schools.

This year also marks the 54th anniversary of the 1964 Civil Rights Act, established as law on July 2, 1964, which was made possible, because the 1954 *Oliver L. Brown et. al. vs. Board of Education of Topeka (Kansas) et. al.* decision struck down the legitimacy of laws that segregated people because of their race.

Lyndon Baines Johnson, the thirty-sixth U.S. President, during the course of his five-year presidency, sounded a death knell to racial inequality through a triumvirate of laws: The Civil Rights Act of 1964, the Voting Rights Act of 1965, and the Fair Housing Act of 1968.

From the social impact emanating from these pieces of historical federal legislation of the U.S. Congress and the LBJ White House, this year also marks the 54th anniversary of The Great Society, first proposed by President Johnson on May 22, 1964, and featuring his presidential pronouncement, "The Great Society," originally crafted by his Harvard speechwriter, Richard ("Dick") Goodwin. LBJ's unprecedented and ambitious domestic vision changed the nation. Half a century later, it continues to define politics and power in America.

President Johnson introduced his vision of a "Great Society" in a May 22, 1964 speech: "The great society rests on abundance and liberty for all. It demands an end to poverty and racial injustice, to which we are totally committed in our time."

The Great Society legislation included "War on Poverty" programs, according to the learning blog of The New York Times, many created under the Economic Opportunity Act of 1964, which established jobs and youth volunteer programs, as well as Head Start, which provided pre-school education for poor children. Johnson's social welfare legislation also consisted of the formation of Medicare and Medicaid, which offered health care services for citizens over 65 and low-income citizens, respectively.

***

Evolution of Institutionalizing the Chief Diversity Officer

The Chief Diversity Officer roles inside institutions (including governments, corporations, universities, and philanthropies) largely self-manifested and shaped themselves in a variety of ways – some reported directly to the CEO, while others reported up through human resources departments. Some organizations created formal departmental infrastructures that tracked and measured success of diversity initiatives led by Chief Diversity Officers.

Other organizations established these diversity czars as independent individual contributors, whose chief messenger role was to persuasively spread the message of diversity across the organization through influence, collaboration and communications. Many diversity chiefs were aligned with external relations, whereby these individuals were often called upon to represent their companies at events that catered to the organization's diverse constituencies, interests and stakeholders.

Efforts to address the issue of diversity both domestically and globally, not only inside multi-national enterprises and institutions, but also across the social fabric, began long before the evolution of the chief diversity officer role, but took shape in other forms, such as affirmative action, as a result of legal and compliance mandates. However, other forward-thinking organizations were also strategically formulating their own positions on diversity and enterprise value.

I know because I witnessed some of those companies' diversity impacts (like Procter and Gamble, for instance, under John Pepper, former Procter and Gamble CEO, and the late LaVelle Bond, former Procter and Gamble Vice President for Diversity Worldwide) inside the Clinton Administration, when I served as a senior policy adviser in the Clinton White House Office of Science and Technology Policy (OSTP). Therein, LaVelle Bond, the late Charles M. Vest, former MIT president, Judith Rodin, then-president of the University of Pennsylvania and now president of the Rockefeller Foundation, Shirley Malcolm, head of human resources at the American Association for the Advancement of Sciences (AAAS), the late John H. Gibbons and Neil F. Lane, former assistants to the president for science and technology policy and OSTP Directors, and so many fine others and I led OSTP's contribution in 1998 to "Clinton's Initiative on Race," which resulted in the White House policy document, *"Meeting America's Needs for the Scientific and Technological Challenges of the Twenty-First Century – A White House Roundtable Dialogue for President Clinton's Initiative on Race."*

African American leaders in business, government, academia, and philanthropy have a unique contribution to make to the challenges faced by African American men and boys, particularly in regard to the development of talent needed for the future sustainability of American society. In response to these challenges, as we stated earlier, Morgan State University and The Kellogg Foundation co-sponsored a dialogue on June 26-27, 2007 at the Lansdowne Resort Boardroom in Lansdowne, Virginia entitled, *Future of African American Men and Boys: Promoting the Saving, Transforming and Empowering of African American Men and Boys for the Betterment of American Society.*

The Lansdowne dialogue was co-chaired by Dr. Calvin O. Butts III, President, SUNY Old Westbury, and Dr. Earl Richardson, then-President, Morgan State University. Butts and Richardson also served as co-moderators of the conversation. This Lansdowne dialogue brought together 12 prominent African American leaders to inform business, government, university, and philanthropic interests on suitable goals and strategies for improving and enhancing the lives of African American men and boys in American society.

The fixes that both extraordinary events put into place for America are far from complete. Scholars and The Obama White House called for a "black male initiative" to focus not on segregation, but achievement, and are basing it on workplace data. I have written what that initiative looks like a little further elsewhere on the *Future of African American Men and Boys*, featured on LinkedIn Pulse *Social Impact* Channel and on my website. But, I think it is one of the most important things business leaders can do to make American business — and society — stronger.

Today, the Chief Diversity Officer positions are much more commonplace and widely accepted across multiple industries, including higher education, government and not-for-profit entities. Most organizations now have high-level diversity officers, whose role is to make their work places more culturally sensitive, and whose role is to ensure business enterprises and corporate brands are viewed as places that strategically manage diversity integrally aligned to enterprise value for shareholders and stakeholders. Awards and recognitions are nowadays given to those institutions that are cited for their innovative diversity initiatives.

As the diversity function evolved over the years, affinity groups sprouted up so that people of similar backgrounds could share experiences and discuss their unique views. Diversity was no longer a "one-size-fits-all" business value proposition. In the age of demography shift and heightened engagement of shareholders and stakeholders, companies now have to consider the perspectives of different affinity groups that emerged around race, gender, sexual orientation, religion, age, physical disability, military status, and so many other dimensions.

More specifically, eight criteria for companies to be listed exclusively inside the prestigious DiversityInc's Top 50 Companies for Global Diversity include:

1. Presence and role of a global diversity council
2. Effective use of global employee resource groups for recruitment and talent development
3. Global policies to prevent discrimination and harassment
4. Global initiatives to hire and promote people with disabilities
5. Global initiatives to hire and promote LGBT people
6. Cross-cultural mentoring initiatives
7. Specific talent- and leadership-development initiatives for women
8. Global supplier-diversity initiatives

According to DiversityInc, "The No. 1 company on this list, Deloitte, has been at the forefront of initiatives to advance women and other underrepresented groups in a variety of countries in Europe, Asia, the Mideast and Africa. Seventy-five percent of its employees work outside of the United States. Global resource groups include those based on gender, age, heritage, religion, disability, sexual orientation and parenting."

Here is the 2014 DiversityInc Top 10 Companies for Global Diversity

1. Deloitte
2. IBM
3. Sodexo
4. EY
5. Accenture
6. Procter & Gamble
7. Merck & Co.
8. Dell

9. Johnson & Johnson
10. Wyndham Worldwide

Photo Credit: <u>Walmart</u> is committed to initiatives that support diversity and inclusion.

***

It begins at the top.

We are still not quite truly diverse in our boardrooms, in our C-suites, in our colleges and university leadership, and in our institutions or other bastions of real power and influence.

<u>BLACK ENTERPRISE</u> reports that "corporate boards have become less diverse over the past several years. The report, *"Power in the Boardroom,"* is also featured as the cover story in the July-August 2014 issue of BLACK ENTERPRISE Magazine.

In its second annual report focused on African American representation of corporate boards, "the media company <u>identifies 176 African American directors at S&P 250 largest companies,</u> including American Express, Walmart, Xerox and Carnival Corporation, on the BLACK ENTERPRISE Registry of Corporate Directors."

Read more inside: Black Enterprise Wealth for Life. The registry and report can be found here. Remarkably, the report "reveals 74 companies with no African American representation among their boards of directors."

According to a report from the *Alliance of Board Diversity*, in 2012, **white men held 75% of board seats** on the 500 largest publicly traded companies, versus **5.5% for African American men** and **1.9% for African American women**.

Rev. Jesse Jackson, founder and president of the civil rights group Rainbow PUSH Coalition, is among those challenging the lack of board diversity, taking aim at Silicon Valley high technology companies, like Google Facing Test on Diversity, and pressing top management on the subject."

> *"By denying African Americans access to a seat at the table, they have made a detestable statement that they seek to maintain these preserves of white male privilege and dominance."- Earl "Butch" Graves Jr., CEO and President of BLACK ENTERPRISE*

***

What have these diversity and inclusion programs accomplished that truly permeates across the social fabric?

With all of the time, infrastructure and resources allocated to making us all more accepting of our differences, here we are nearly 15 years after the start of the millennium with racial tensions, LGBT issues, women's rights, pay equity, economic disparities, political divisiveness, voting rights and civil rights dominating our daily lives. There seems to be a rising plethora of racially charged incidents of late coming from multiple segments of our society.

A Harvard Business Review blog argues that diversity training can promote prejudice.

"Diversity training doesn't extinguish prejudice. It promotes it."

The blog, citing a study of 829 companies over 31 years, showed that diversity training had "no positive effects in the average workplace." Millions of dollars a year were spent on the training resulting in, well, nothing. Attitudes — and the diversity of the organizations — remained the same.

It gets worse. The researchers — Frank Dobbin of Harvard, Alexandra Kalev of Berkeley, and Erin Kelly of the University of Minnesota — concluded that "In firms where training is mandatory or emphasizes the threat of lawsuits, training actually has negative effects on management diversity."

The solution proposed by the Harvard Business Review for the divisiveness of diversity initiatives is rather than engaging people through the lens of race, gender, age, heritage, religion, disability, sexual orientation and parenting, we need to engage people as people.

"Stop training people to be more accepting of diversity. It's too conceptual, and it doesn't work," says the Harvard Business Review. "Instead, train them to do their work with a diverse set of individuals. Not categories of people. [Just as] People."

"Teach them how to have difficult conversations with a range of individuals. Teach them how to manage the variety of employees who report to them. Teach them how to develop the skills of their various employees," argues the Harvard Business Review.

"Move beyond similarity and diversity to individuality."

At its core, "diversity", as it is used in relation to the workplace, is a divisive and rather weird concept, reports The Guardian (U.K.). "In claiming certain groups into its fold, it suggests that some people are "diverse" and some are "not diverse". It suggests, in other words, a nucleus of normal and goes about classifying everyone off-centre into check-box categories that can be totted [or totaled] up and turned into tables for the annual report."

"What's more, definitions of diversity tend to be skin-deep, about differences you can see [...]," The Guardian (U.K.) concludes.

It appears that whenever a high profile and potentially racially divisive incident occurs, the battle lines are drawn. Even if the incident or issue (albeit health, human services, housing, education, energy, sports, entertainment or environment) itself has nothing to do with race, it quickly turns into a racial issue, when racial stereotypes surrounding those involved come into play.

© AP

According to <u>The Washington Post</u> on November 5, 2008, in a historic commemorative article entitled, "Obama Makes History – U.S. Decisively Elects First Black President, Democrats Expand Control of Congress" – a piece now framed and bronzed in millions of black family homes, "standing before a crowd of more than 125,000 people, who had waited for hours at Chicago's Grant Park, Obama acknowledged the accomplishment and the dreams of his supporters," as they ushered in the Obama era – the new Age of Obama.

"The historic Election Day brought millions of new and sometimes tearful voters, who had waited in long lines at polling places nationwide, and celebrations on street corners and in front of the White House. It ushered in a new era of Democratic dominance [...] and returning them to a position of power that predates the 1994 Republican Revolution [...], according to the historical November 5, 2008 election special commemorative edition of The Washington Post.

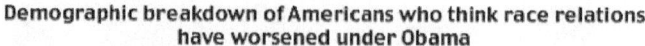

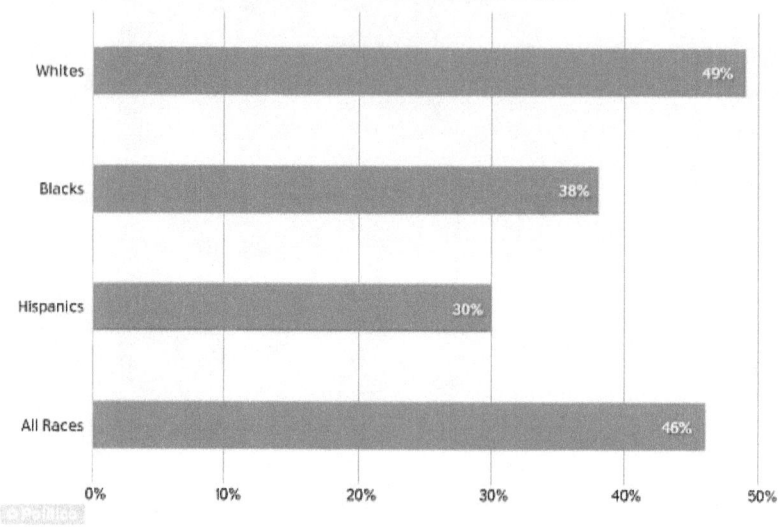

Demographic breakdown of Americans who think race relations have worsened under Obama

Whites — 49%
Blacks — 38%
Hispanics — 30%
All Races — 46%

Photo Credit: Politico *and* The Daily Mail (U.K.)

For a brief moment, we thought the election of our first black president was a signal that we had turned the corner and become a more racially neutral country. On the contrary, President Obama's election appears to have added fuel to the fire. Nearly half of Americans in key voting areas think race relations have gotten worse under the nation's first black president, a new poll shows, reports Politico and The Daily Mail (U.K.).

The survey, taken by Politico the last week of August and the first week of September, found that 46 percent of residents in states and House districts with competitive federal elections this fall believe that racial tension has increased under President Barack Obama's leadership.

Just six percent of voters polled said they thought relations had improved under Obama. The survey's findings follow the high-profile death of the black teen, Michael Brown, in

Ferguson, Missouri, who was shot and killed by a white police officer, reports The Daily Mail (U.K.).

America was neither ready for its first black president, nor was its first black president ready for America. We are not racially neutral and it was the election of President Obama that fueled this tipping point. Blacks praised this victory and dared anyone to take issue with this president, at his height of minimum political risk in 2008-09, because all criticism was viewed as an attack on race, and not an attack on policy in the national interest.

"There has been quite an interesting response to President Obama's second term selections for filling top cabinet positions. Women, such as Hillary Clinton, have stepped down from their cabinet positions and President Obama has nominated only men to fill the vacated positions [...], even though there are qualified women who could have taken the job." points out political blogger, Brian Anderson in early January 2013.

"The argument for all male nominations is that, "well, these were just the best people for the job. We don't look at gender, we look at qualifications." This logic makes certain women advocacy groups angry at the suggestion that there was not even a single woman as qualified as one of the men he has nominated," Anderson argues.

Anderson writes further, "You don't put women in top cabinet positions just because you feel like you have to be politically correct, but because diversity is extremely important when it comes to making important decisions. You want people from a variety of backgrounds and upbringings when considering policy. That keeps you from overlooking details that may not adversely affect people that look like the decision makers, but may very well hurt those not involved in the decision making process."

"The opinions of women especially should be sought after to bring a different and much needed perspective in making such important policy decisions." Anderson adds.

President Obama's election presented an unprecedented opportunity for the "disenfranchised white male" to come out too. This disenfranchised group was already bearing the brunt of the corporate diversity initiatives that had now raised the sensitivity that rooms full of "white males" were a sure sign that an organization lacked diversity in its strategic thinking and management.

It was more rare for white males to even be considered for chief diversity officer roles, although now we are beginning to see this co-hort represented. For those who did not really buy into all of this "talk about diversity," we now have the Rush Limbaugh's, Sean Hannity's, Mark Levin's, Michael Savage's, Breitbart's out there, who could now voice either satisfactions or frustrations we have been feeling for years, as we are forced to accept these so-called "diversity initiatives."

While racial undertones have always been a part of our society, we are seeing much more blatant acts now being exposed. Instead of talking more about ways to celebrate our differences and to appreciate our diversity, we are seeing more and more negativity being expressed.

Did putting diversity in the spotlight help fuel this negative exposure?

We are now a more socially-connected society through our hand-held mobile devices, making it much easier to spread opinions and views.

The relative anonymity in which we can communicate within social media provides a way to make raw, blatant comments that we would most likely be reluctant to make face-to-face. We seem to be speaking to each other more exclusively nowadays inside the news and opinions we spread among each other instantaneously on our digital hand-held devices.

As ESPN's Ivan Maisel recently reported on Saturday, September 20, 2014, Florida State quarterback, Jameis Winston "picked the wrong week to shout something demeaning to women."

"If this is the nation's Rosa Parks moment regarding violence toward women, Winston picked the wrong week to mimic the worst form of fratboy behavior at the top of his lungs, while atop a table at the student union," Maisel writes.

Florida State deserves credit for suspending Winston for the entire game against ACC Atlantic rival Clemson one day after the incident.

We all tend to speak over each other and not with each other so fast that conventional wisdom spreads with such exponential virility, like a brush fire, such that no containment or quarantine of information, knowledge or understanding is possible.

We have seen countless instances that demonstrate that we are much less tolerant of others, much more outspoken in our politically incorrect views, and with very little to show for all of the dollars spent on conferences and programs aimed at making us more diverse and more racial and gender tolerant.

***

Bottom-Line Takeaway:

So, here we are today – more distrusting of others, and more closeted in our views. For diversity initiatives to work, they must allow us to discuss our true feelings and biases, and not be chastised for what we believe.

We must evolve as a society where we can relegate those who harbor views of what can be described as racist, to the commonly viewed and reasonable point of distaste or disdain. Yet, along the way on the course to this new destination of societal norms and conventions, what have we done to understand why we feel this way?

Take the explosive case of Paula Dean. She honestly answered a question that was asked of her. Yet, she was slaughtered in the press alongside her food empire being attacked. Blacks surprisingly came to her defense. For them, what she said and did was troubling of course. In her own, southern charming way, this woman, who grew up in the height of racial segregation, simply told the truth. Some could say she was too naïve to know better. However, her naivety is what endears her to us. She admitted what she said and felt. Since then, she has taken steps to face her inbred prejudices.

Contrast the case of Paula Dean's comments, to what takes place in corporate offices, where discussions about the racial makeup of the leadership teams surely take place.

Some would say, "We have become a very diverse nation and diversity, due to its very nature, breeds disagreement. People have always had trouble getting along with each other, but in our day we find ourselves in a divided country."

Photo courtesy of Shutterstock.

In part, "we are a nation divided because of two things which are mutually exclusive – liberty and government. While some people seek a government that passes binding laws that infringe on personal freedom, others seek a more libertarian form of government. While one group sees the government as the solution to our problems, another sees it as the cause of our problems," some would add.

Yes, America, we have a long way to go before we truly accept each other and our differences. We need to learn how to fully appreciate the rainbow of colors, ideas, lifestyles and philosophies in each other. We must learn not to judge others, because they hold views and opinions quite dissimilar to ours. It is only then will we be a society that truly appreciates our differences, and values those perspectives that we all have.

**

Sidebar: Planning to Give Forward to HBCUs

Baby-boomer retirees plan to set new standards of diversity and participation in giving forward to historically black colleges and universities (HBCUs) during commencement season.

Baby-boomer generational philanthropy and charitable giving is raising the bar for future generational giving forward.

Baby-boomer retirees say, "I want to have an effect as a supporter" as an alumnus at my annual HBCUs commencement gala!

Watergate's impact on institutions during the baby-boomer era inspires a need for integrity and trust held by these donors in such well-endowed institutions, including churches, K-12 schools, HBCUs and related charities.

Billionaire investor and philanthropist Robert Smith made a $40 million dollar pledged to eliminate student debt for the entire 2019 graduating class at Morehouse College.

Smith said, "I know my class of 2019 will make sure they pay this forward ... and let's make sure every class has the same opportunity going forward."

Smith is the first African American to sign the "*Giving Pledge*," a campaign started by Bill and Melinda Gates and Warren Buffett that encourages the world's wealthiest individuals to commit to dedicating the majority of their wealth to philanthropic causes.

In signing the pledge, Smith wrote, "Potential is no guarantee of progress. We will only grasp the staggering potential of our time if we create on ramps that empower ALL people to participate, regardless of background, country of origin, religious practice, gender, or color of skin."

Smith continued, "My story would only be possible in America, and it is incumbent on all of us to pay this inheritance forward."

Inspired by baby-boomers' inherited responsibility and accountability in giving forward to HBCUs, Generation-X givers declare "such well-endowed institutions are participating together in doing good things in education capacity-building" creatively through more diverse charitable options, venues and opportunities.

Generation-X givers also want to make some impact giving forward chiefly in prevailing societal healthcare concerns of cancer research, oncology patient care, population-based clinical services, and high-end specialized holistic tertiary business care, including transplant, intensive-care and trauma services.

Contrast this going forward to "millennial generation" givers, who say "it is essential we support well-endowed HBCUs in diverse ways. And, our multicultural, multi-ethnic world needs to fully participate during commencement season and homecoming galas in giving forward to HBCUs right away."

Draw an additional distinction and look even further forward to "21st Tercentennial generation" givers. They participate, engage and trust such well-endowed HBCU institutions to "make a difference." Cohorts of this highly-shifted demography of givers have high expectations of a diverse HBCU institution working for future benefits of the next generation and the generation after that. These donors say, "Show Us," and "Prove It, Now!"

Moreover, these 21st Tercentennial generation givers say, "We are looking way far out ahead in our giving forward to HBCUs." This cohort of givers are even reaching as far out and beyond towards America's Tercentennial – *On Getting to 2076!*

Altogether, humanitarian and moral benefits to baby-boomer donors and their private retirement philanthropic endeavors have been the central motivation and intent of baby-boomer giving forward planning to HBCUs. Tax benefits resulting from such good societal endeavors on the road to stakeholder value have also assumed increasing importance.

For the first time ever, charitable giving exceeded the $400 billion mark in 2017, spurred by growth from all four sources of giving.

$410.02 billion

Where did the generosity come from?

Giving by Individuals
$286.65 billion
5.2% ▲ 70%
increased 5.2 percent (3.0 percent when inflation-adjusted) over 2016

Giving by Foundations
$66.90 billion
6.0% ▲ 16%
increased 6.0 percent (3.8 percent when inflation-adjusted) over 2016

Giving by Bequest
$35.70 billion
2.3% ▲ 9%
increased 2.3 percent (0.2 percent when inflation-adjusted) over 2016

Contributions by source
(by percentage of the total)

Giving by individuals increased $14.27 billion over last year for a growth rate of 5.2 percent, on track with the growth in total giving.

This has become especially acute, as the complex institutional advancement, academic and research capacity-building needs of HBCUs have substantially grown in scale and scope now and into the future.

BRIDGING THE BLACK RESEARCH GAP

On Integrated Academic and Research Capacity Building at Historically Black Colleges and Universities

Oliver G. McGee, Ph.D.

America is a generous nation.

Remarkably, Americans are an extraordinarily generous people. Most of all, our generosity is not necessarily just tied to tax benefits of giving forward to HBCUs.

Alexis de Tocqueville in "Democracy in America" records the American citizenry gives forward "without reference to any bureaucracy, or any official agency."

Inside "Giving USA 2018", published by Indiana University's Lilly Family School of Philanthropy, Americans are indeed extremely generous in giving forward to charities and philanthropic organizations at $410.02 billion in 2017, crossing for the first time the $400 billion-dollar threshold fueled by a bullish stock market and 3.2 percent U.S. economic growth. This is equivalent to the net income of about 20% of all corporations of the Fortune 500 today.

To achieve this exceptional $410 billion amount of charitable generosity experienced by our citizenry (including baby-boomer retirees, and the largest emerging millennial generation — which is about an 8% larger cohort than the baby-boomers), America invested an equivalency of nearly $4 trillion dollars in assets, sold about $6 trillion dollars in goods and services, and drew upon the combined productivity of more than 40 million employed workers under record low 3.6% unemployment.

That's a lot of goodwill for all charitable Americans to feel good about giving forward to HBCUs, for sure!

———————

Former Clinton White House Senior Science Policy Advisor and former U.S. Deputy Assistant Secretary of Transportation gives his take on baby-boomer philanthropy and charitable giving forward to the nation's HBCUs.

Oliver McGee, PhD, MBA, 2012-13 American Council on Education Fellow at UCLA, says "planning to give forward in serving HBCUs, is 'making a difference' and feeling good about it." He is certified in fundraising management at Indiana University's School of Philanthropy in 2013. He is a 2013 NASDAQ Certified Director in Directorship and Corporate Governance from the 26th Director Education and Certification Program of the UCLA John E. Anderson Graduate School of Management.

Table of Contents

Dedication i

Acknowledgements ii

Foreword v

Appendix: 1999 Proceedings of Panel Discussion and Position Papers, *Meeting America's Needs for the Scientific and Technological Challenges of the Twenty-First Century, A White House Roundtable Dialogue for President Clinton's Initiative on Race*..1

1999 Letter of Assistant to the President for Science & Technology...2

Contents...4

About The Panel: OSTP-AAAS *One America Conversation*......7

Summary of *One America Conversation*..........................9

Fact Sheet: National Science and Technology Council (NSTC) 120

Fact Sheet: President's Committee of Advisors on Science and Technology (PCAST)...126

About the Editor...134

Appendix

Meeting America's Needs for the Scientific and Technological Challenges of the Twenty-First Century

A White House Roundtable Dialogue for President Clinton's Initiative on Race

Proceedings of Panel Discussion and Position Papers

150th Anniversary Meeting of the
American Association for the Advancement of Science
Philadelphia Marriott Hotel
February 13, 1998
2:00-4:30 p.m.

"First, science and its benefits must be delivered toward making life better for all Americans – never just a privileged few ... Science must not create a new line of separation between the haves and the have-nots, those with and those without the tools and understanding to learn and use technology ... Science can serve the values and interests of all Americans, but only if all Americans are given a chance to participate in science."

President William Jefferson Clinton

January 1999
Executive Office of the President
Office of Science and Technology Policy

THE WHITE HOUSE

WASHINGTON

January 7, 1999

Dear Colleague:

On June 13, 1997, President Clinton issued Executive Order No. 13050, which created the Initiative on Race and authorized the creation of an Advisory Board to advise the President on how to build one America for the 21st Century. The Board was tasked with examining race, racism, and the potential for racial reconciliation in America using a process of study, constructive dialogue, and action. The Board also focused on the role race plays in civil rights enforcement, education, poverty, employ-ment, housing, stereotyping, and the administration of justice, health care, and immigration.

I believe scientists and engineers have a unique contribution to make to the challenges faced by the President's Initiative on Race, particularly in regards to the development of talent among groups tradi-tionally underrepresented in science, mathematics, and engineering. Therefore, I am pleased to release the proceedings of the panel session on *Meeting America's Needs for the Scientific and Technological Challeng-es of the Twenty-First Century*, held February 13, 1998 at the 150th Anniversary Meeting of the American Association for the Advancement of Science (AAAS). This panel session helped inform the Administration on suitable goals and strategies for diversifying the scientific and techno-logical community. The richness of this input is rooted in the experience and diversity of the participants, who came to the panel session from academia, industry, laboratories, professional societies, and government.

The panel session's co-chairs, Dr. David Hamburg, President Emeritus of the Carnegie Corporation of New York, and Dr. John H. Gibbons, then-Assistant to the President for Science and Technology are to be commended for their efforts in pulling together a truly distinguished panel and an enormously informative conversation on the President's goal of *One America*. A diverse scientific and technological workforce will ultimately position the United States for continued leadership in the coming millennium.

Sincerely,

Neal Lane

Assistant to the President

for Science and Technology

CONTENTS

Letter from the Director of OSTP..

About the Panel: OSTP-AAAS *One America Conversation* ....

Summary of OSTP-AAAS *One America Conversation* ....

    Opening Remarks ....
    Panel Discussion ....
        Need to Diversify the S&T Community ....
        Education Access for Minority Children in S&T...........
        Role of Mentoring ....
        Outreach Programs that Work ....
        Target Minority Recruitment in Science
            and Engineering ....
        Educational Barriers ....
        Race and the Role of Standardized Tests ....
    Closing Remarks ....

Appendix A *Central Goals of One America in the 21st Century:*
The President's Initiative on Race..

Appendix B *White House Letters to Invited Panelists and Contributors*...

Appendix C *Panel Agenda: OSTP-AAAS One America Conversation*......

Appendix D *Biographical Summaries of Invited Panelists and Contributors*..

Appendix E *Position Papers of Invited Panelists*....................................

    The Advancement of Science
        Shirley Malcom..
    Partnering for Workforce Development: A Model for
    Increasing the Supply of Skilled Workers

Meeting America's Needs for the Scientific and Technological   5
   Challenges of the Twenty-First Century – A Retrospective
       Cathleen Barton...
Development Fund for Black Students in Science
    and Technology
     Julian Earls...
Under-representation Perspectives from Academia
     Carlos Castillo-Chavez..................................
Diversifying the Science and Technology Community
     Richard Tapia..
Race & the Misuse of Standardized Tests in Predicting
    Academic Potential
     Percy Pierre..
Achieving a Diverse Science and Technology Community
     Samuel Massie...
On the Retention of Under-represented Minorities
    in Science
     Lydia Villa-Komaroff...................................
Absence of Minorities From Research Fields Will Result
    in Grave Consequences in U.S.
     John Aderete...
Why America is Still "A Nation at Risk" Fifteen
    Years Later?
     Darleane Hoffman.......................................
Addressing the Issue of Under-representation of Minority
    Groups in Graduate Engineering and Science
    Education
     Howard Adams...
Personal Reflections on Race and Achieving S&T Diversity
     Satya Atluri..
Beyond Conflict or Compromise
     Fred Begay...
Procter & Gamble's Goals and Perspective
     O. LaVelle Bond...
Workforce Diversity in the S&T Community:
    Key Strategies & Future Directions
     Herbert Wong..

Appendix F *Position Papers of Invited Contributors*...........................

Engineering and Affirmative Action: Crisis in the Making
   George Campbell, Jr....
Race and Fear: The Real Hot Buttons Behind the Diversity Debate
   William Gray, III...
Are We Still A Land of Opportunity?
   Charles Vest...

ABOUT THE PANEL:
OSTP-AAAS *ONE AMERICA* CONVERSATION

Science and technology offer exciting opportunities to shape America's future. Yet, science and technology fields traditionally have attracted a less diverse pool than many of the non-technical fields. In the 1945 report to President Harry S. Truman defining a national program for post-World War II scientific research, *Science-The Endless Frontier,* Vannevar Bush reported that "There are talented individuals in every segment of the population, but with few exceptions those without the means of buying higher education go without it. Here is a tremendous waste of the greatest resource of a nation – the intelligence of its citizens. If ability, and not the circumstance of family fortune, is made to determine who shall receive higher education in science, then we shall be assured of constantly improving quality at every level of scientific activity."

As part of its determination to propel the Nation into the 21st century on a strong scientific and technological foundation, the Clinton Administration articulated several goals in the policy document, *Science in the National Interest.* Two of the major goals are the production of the finest scientists and engineers for the 21st century and scientific literacy for all. To help achieve these goals, the Administration is committed to maximizing the Nation's pool of talented, well-educated, and highly trained scientists and engineers. This entails maintaining demonstrated excellence in the production of scientists and engineers, by actively increasing the participation of talent reflective of the Nation's diversity.

Scientists and engineers have a unique contribution to make to the challenges faced by the President's Initiative on Race (see Appendix A), particularly in regards to the development of talent among groups traditionally underrepresented in science, mathematics, and engineering. In response to these challenges, the Office of Science and Technology Policy (OSTP) and the Directorate for Education and Human Resources Programs of the American Association for the Advancement of Science (AAAS) co-sponsored a *One America* panel session at the AAAS 150th

Anniversary Meeting entitled, *Meeting America's Needs for the Scientific and Technological Challenges of the Twenty-First Century*. The OSTP-AAAS *One America* panel was co-chaired by The Honorable John H. Gibbons, then-Assistant to the President for Science and Technology, and David Hamburg, President Emeritus of the Carnegie Corporation of New York and a member of the President's Committee of Advisors on Science and Technology (PCAST). Hamburg also served as the moderator of the conversation. The panel session brought together twenty prominent scientists and technological leaders of academia, industry, and government to discuss the need for participation of all Americans in science and technology, and the responsibility and expectations of a diverse scientific and technological community in contributing through research and education to the national good. Biographical summaries of the invited panelists are given in the Appendix A. What follows is a brief summary of the OSTP-AAAS *One America* panel discussion.

SUMMARY OF
OSTP-AAAS *ONE AMERICA* CONVERSATION

The OSTP-AAAS *One America* panel sought to carry forward a discussion of the ideals that underlie a government-academe-industry partnership for creating a diverse scientific and technological community, and it was organized around the basic goals of the President's Initiative on Race. The discussion focused on three issues: the need to diversify the science and technology community; how to encourage minority students to choose technical careers; and how to deal with the challenges to recruitment programs in science and engineering that target minorities.

***

Opening Remarks

John Gibbons convened the panel discussion, and offered in his opening remarks the purpose of the session with respect to President Clinton's goals through his Initiative on Race, and to science and technology policy. Gibbons stated, "in the President's Initiative on Race, we have an effort to move the country much closer to a stronger, more just, and a more unified nation, one that offers opportunities and fairness for all Americans. It is a chance for every citizen in our country to be part of a national conversation about racial diversity in America, and about the strength that diversity brings to our nation, and indeed, to parts of the nation like the science and technology community." America needs more workers trained in science and technology. The global economy, which has been created by advances in transportation and communication technologies, has broadened the distances between the "haves" and "have -nots." We have to carefully craft our application of these advances, so that we can narrow, rather than broaden, the gaps between our socioeconomic communities. Gibbons urged that we in science and technology have a unique contribution to make to these challenges, and the kinds of challenges that the President has put before us in his Initiative on Race.

The moderator for the panel discussion, **David Hamburg**, discussed the common themes flowing from the position papers, which were submitted by the panelists and invited contributors prior to convening for the conversation. Hamburg spoke of the serious barriers against minority participation in science and technology that must be addressed. He described the need for ensuring that all Americans can understand issues that have technical content in the public arena. Hamburg also researched the issue of the demographic changes that can be foreseen in the next century. He stated they are "so far reaching that they call for a re-assessment of the opportunities to diversify science and technology achievement in education and career opportunities in this country."

Hamburg suggested that we can enlarge the scope of opportunity for distribution of knowledge and skills in the population, and that the challenge is for public policy to highlight the public-private best practices and implement them on the national scale. Finally, Hamburg urged that we must live up to the promise of life-span development of people from childhood to adulthood by encouraging a higher level of learning through science and technology.

***

Panel Discussion

Need to Diversify the S&T Community

The under-representation of minorities in science and technology raises important concerns regarding both equal opportunity and the future ability of the nation to produce an adequate number of scientists and engineers. Demographic data show a workforce increasingly comprised of African Americans, Latinos, Asian Americans, Native Americans, and persons with disabilities - groups historically underrepresented in science, mathematics, engineering, and technology. This country must invest in education and training for these groups not only to make good on its promise of equal opportunity, but also to insure that employers have a qualified workforce in the 21st century. **John Alderete** drew attention to the growing demographic shifts in the U.S. population in the next century. He cited that, "in just two or three years, if not sooner, Latinos will be the majority minority population in this country, and that in less than 50 years, there will be 100 million Latinos in this country - the single largest minority this continent has ever seen." Alderete noted that "only five percent of us will be elderly, and eighty-five percent of us will have been born here multi-generational." He further warned that "if you think the 1990 LA riots were bad, we haven't seen anything yet, if we don't start educating more people."

Carlos Castillo-Chavez and **Richard Tapia** agreed that we still have serious barriers against minority participation in science and technology which must be addressed. According to Tapia, "under-representation in science and technology is not just a health of science issue, it's a health of the nation issue." **Herbert Wong** urged that "we really need to learn how to leverage the differences, as well as, the similarities amongst ourselves, and we must not enter into a competitive or conflictual area around our differences. We need to figure out how to draw value from our diversity and to create knowledge in order to gain from our differences." Wong added that "effective change comes because of institutional and organizational change, rather than just individual or small group efforts. We can make some progress through our individual energies and efforts, but we really need to mobilize this effort into institutions and organizational frameworks." Wong further suggested that "we have got to build workforce environments of inclusion. We have to pay attention to social support of workers. We have to build those healing communities that address the across-group concerns and conflicts. Without this effort, knowledge creation and productivity is severely diminished."

Satya Atluri pointed out that "as intellectual property becomes increasingly global, most of these international students are exporting knowledge back to their homelands, and I think this is a serious tragedy for the country." Atluri urged that "we should work on national policies that encourage persons of all races or national origin to thrive in this country and open all the doors to them at all levels." Atluri would also like to see the country adopt the concept of magnet institutions - perhaps

10 or 15 institutions - where massive amounts of resources are synergized for minority students, especially for those who are native-born.

Educational Access for Minority Children in S&T

Access to education is, above all else, a promise to all children. We need to take full advantage of emerging computer and communication technologies to meet this promise. The panelists acknowledged the Administration's commitment to expanding these technologies to public schools, and work to connect every classroom in the country to the Information Superhighway. They also recognized President Clinton's concern for the need to expand our capacity to conduct research on education - what works, what doesn't, and how new approaches to education can be successfully implemented in America's classrooms.

Still, the panelist felt that the only sure way we can help every child in America to climb the "ladder of opportunity" is through increased Federal funding to improve math and science education for all children and to increase access to job training in technical fields. The panelists urged that this is essential in a new economy forged by expanded trade and technology, and constrained by skilled worker shortages. If America's high school children are graduating with math and science competencies well below the competencies of our global competitors, how can we expect them to contribute to America's economic future? All of the panelists agreed that we need to take bold new steps now to reverse the downward trend in math and science education of America's children - especially for African American, Latinos, and Native American children. Various forms of Federal aid to education supported by this Administration were acknowledged as working to facilitate the development and application of standards, to assist school districts serving poor communities, to assure safe and drug-free classrooms, to improve curricula in accordance with changes in our economy, and to empower adult learners entering school for a lifetime. However, the panelists unanimously agreed that we need many more Federal dollars funding grass-roots efforts that show more school children from all communities how to say, "Yes, I can do this."

Role of Mentoring

Shirley Malcom opened the panel discussion by focusing on the tremendous value of mentors and role models for minority participation in graduate education at colleges and universities categorized as Research I institutions. Malcom stated that the AAAS surveyed Research I institutions exclusively in order to ascertain the state of their effectiveness in this area, because they are the producers of future scientists and engineers holding the elite Ph.Ds, who in turn become future faculty at Research I universities, future governmental and industrial policy makers, future PCAST members, and future directors of NSF and NIH. She spoke of the "isolated" educational environment encountered by African American, Latino, and Native American students at Research I institutions, and that such an environment is resulting in a decrease in the enrollment of minority students in the graduate schools of these institutions.

Howard Adams cited that many minority students simply don't receive adequate funding from inside the institution. "If you don't have inside funding, you don't have a key to the lab," Adams cautioned, "You don't have a mentor. You don't have an advisor. You don't have colleagues. You don't belong to the journal clubs. You don't belong to any of the scholarly activities that one must do if you are going to graduate." Adams further warned, "if minority students don't have the advising and the nurturing, they don't know the milestones that will enable them to be successful." Even after we do all of this, Adams urged that "we have to give minority students access to the recommendations (sponsorships) that are going to launch them forward to the next level."

The issue of science and technology mentoring - recruitment, retention, and academic development - was cited by many of the panelist as primary concerns that educators face with minority engineering students. **Richard Tapia**, recalled as a poor, Mexican American child how he excelled in math and science. But, how his technical abilities could lead him out of inner city Los Angeles seemed uncertain. "I just needed someone to tell me, 'Yes, it's possible.'" Tapia recalled several teachers who went the extra mile and encouraged him to "take his talent and love of math, and make a career out of it."

The importance of role modeling and mentoring recognized at the highest level of the Federal government was warmly praised by several of the panelists. The annual Presidential Awards for Excellence in Science, Mathematics, and Engineering Mentoring were established in 1996. Since then, dozens of individuals (including panelists **Tapia, Adams,** and **Carlos Castillo-Chavez**) along with numerous organizations and institutions have been honored for their outstanding mentoring efforts that have encouraged significant numbers of minorities, women, and disabled persons to succeed in these fields. Tapia, Adams, and Castillo-Chavez identified the President Mentoring Awards as a step in the right direction, because it encourages academia to recognize the efforts that many faculty are realizing in the mentoring and nurturing of minority student participation in science and technology.

Outreach Programs That Work

According to **Howard Adams**, "there still remains a lack of information, research, and professional training on establishing effective formalized mentoring programs. Methods and techniques currently being used are disjointed and frequently without guidelines for implementation. The formation of mutual mentoring relationships between the participants (mentors and protégés) are often left to chance." "This hit-and-miss approach," says Adams, "fails to take advantage of proven strategies for the development and implementation of effective mentoring programs." To help fill this void, the National Consortium for Graduate Degrees for Minorities in Engineering and Science, Inc. has established the National Institute on Mentoring (NIM), headed by Adams at the Georgia Institute of Technology to improve existing mentoring techniques employed by universities, industries, and governmental agencies, and to offer consulting services to these same institutions and agencies in the areas of developing and implementing formalized mentoring programs where none presently exist. Adams, like many of the panelists, recommended that more programmatic models like NIM are needed.

Cathleen Barton and **Julian Earls** picked up on the theme of best practices and programs that focus on workforce development and minority community outreach from the industrial and Federal agency perspective. Cathleen Barton spoke on behalf of her responsibility in leading an effort to increase the supply of technically skilled workers in the semiconductor industry. Barton stated that the key approach to developing a technically-trained workforce for the semi-conductor industry is increasing the awareness of parents, teachers, guidance counselors, and re-training adults about the educational opportunities available to give them entrance into the fields of high technology.

Earls, along with a group of black scientist and engineers, decided to take personal responsibility in addressing the problem of financial support for black students who desired to pursue a career in science and technology. Earls described the Development Fund for Black Students in Science and Technology, a Washington, D.C. non-profit organization. He also articulated the NASA position that "it is too late to go into a high school and talk to students about majoring in science and engineering." He also summarized a sample of the NASA research grant programs targeted to historically black colleges and universities, Hispanic-serving institutions, and tribal colleges and universities.

For 160 years, Procter & Gamble (P&G) has been in the business of promoting personnel solely from within. **O. LaVelle Bond** stated that "if we [P&G] don't get the kind of people that we need, we don't get the kind of people to help our business run." P&G is both a marketing and high technology company. Sixty percent of P&G's managerial employees have technical degrees. "By increasing our diversity efforts over the last five years, we have significantly expanded our minority managerial recruits," cited Bond.

To get this done, Bond reported that P&G had to develop business strategies aimed at utilizing the best sources of the brightest talent out there and forming strategic alliances with the communities in which the talent pool resides. "We focus on a handful of models that work out there, and we figure out how to appropriately adopt or adapt them. We do not focus on trying to find 'a silver bullet' to solve this issue of workforce diversity," according to Bond. He further reported that P&G is involved in youth development programs like Inroads, and has contributed nearly $2 million to the United Negro College Fund. P&G alliances with these organizations influence people to make good career decisions, and this has been a very important company objective. P&G has identified about 40 colleges and universities as primary talent sources to meet their workforce needs. Each of these universities is involved with a team that is headed by a corporate officer at P&G. Each of these teams is charged to identify the issues, opportunities, and needs of the recruits and the company.

Targeted Minority Recruitment in Science and Engineering

Regarding targeted minority recruitment in science and engineering, a consensus emerged during the conversation that diversity in science and engineering is crucial to national wealth-building and knowledge-creation. However, America's efforts to achieve this goal are currently stalled in the political and legal aspects of affirmative action. The national debate on affirmative action notwithstanding, in American higher education "...race-based scholarships, where race is the sole determining factor, account for less than 1 percent of all graduate and undergraduate scholarships, and about 5 percent of all scholarships exclusively at the undergraduate level. Scholarships that include race as a component account for less than 5 percent of all scholarships in America," according to William H. Gray III, President of the United Negro College Fund. He further suggests that higher education leaders must "...help the people of the academy understand the need for inclusiveness, and that being inclusive does not mean lowering standards." Most of the panelists agreed that we must provide similar research and statistical data that show the economic and technological competitiveness impact of under-representation in science and engineering.

Howard Adams pointed out that "there are more underrepresented minority students taking mathematics and good science classes than ever before." Adams reported that "there are more Native Americans, Hispanic, and African American students taking the SATs than ever before." "There are more of our kids who want to go to college," according to Adams, "so, the good news is that we are making some difference in this area, although we are not recruiting nearly enough students."

Adams warned that "we still sell graduate education as bad medicine for minority students." He advised that we have to get the appropriate information to minority students about the benefits of graduate school.

Lydia Villa-Komaroff articulated the value of a Ph.D. as a degree of opportunity for minority youth. She thinks all too often the message we send to minority youth is that they shouldn't get a Ph.D. because they won't get a job. In fact, what we really mean is that "it is going to be hard to get a job in academia." However, the truth is that professionals holding Ph.Ds have a significantly lower level of unemployment than professionals who don't, according to Villa-Komaroff. She stated that we, in academia, are far too narrow in our definition of success for our students, particularly for minorities, and we must address this.

Educational Barriers

"All of our children can learn, but many of us who teach math and science don't really give our children a chance to learn," cautioned **Samuel Massie**. He further suggest that "as we teach our children, we look at newer methods of presenting math and science material - some-how we have got to bring it down to a level that a child can understand."

James Comer believes, "we have to give a lot of attention to pre-kindergarten, elementary and high school teacher training. The way we teach just misses the way young people learn. So, we have to address the way the education colleges are really preparing teachers to teach...based on the way children learn."

Comer further suggested that we must encourage our children towards a higher level of learning and thinking. He cited several examples to support this. "Children are born explorers, investigators, and scientists," Comer acknowledged. "We have to encourage parents and teachers to nurture that curiosity, to help teachers learn to tap into a child's curiosity, talents, and interests. And hopefully, we can encourage their interest in science and technology."

Darleane Hoffman urged that "we have to start introducing science at the kindergarten level." She suggested, "we can study things like buoyancy. We can teach our children in the bathtub, showing them how things float and what their densities are, and so forth." She urged teachers and parents to develop simple activities using a graded approach. More important, "we should take the view that not everybody is going to be a Ph.D. scientist, nor should they be, any more than all of us should be musicians or artists or sports personalities." Hoffman urged that we should appeal to children at all levels, and show them how science impacts their daily lives. Hoffman believes that eventually, those children who really want to pursue science as a career will come, and that those youngsters who have other primary interests can still find something which will lead them to the scientific method of thinking. Hoffman has become very convinced that "we have to start very early, that we have to take a graded approach, and that we need the teacher training. I hope that in the next 15 years we can make more progress than we have in the last 15."

In the area of bilingual education, **John Alderete** recommended that we must explore critical areas of new research on the language learning skills of all children. He warned that we need more research to determine if bilingual education is working, and if it is the best way to approach the education of Hispanic children. Alderete also called for more investment in bilingual instruction and curriculum development to give children in under-served groups greater access to basic math and science skills using their native language, so they can eventually make the crossover to higher technical learning using English. He argued that learning the fundamentals of math and science is difficult enough without having to overcome the additional barriers of language and access to feedback. Alderete challenged, "Scientists must provide the data, so policies can be altered."

Fred Begay cited a critical problem faced by Native Americans is the absence of sustained Federal funding for education. Begay recom-mended that the National Science and Technology Council devise some strategy to support K-12 science and math education on a long-term basis, especially in the tribal communities. Begay acknowledged that "NSF will come in and support us for a few years, and then they are gone."

Race and the Role of Standardized Tests

The role of standardized tests was identified by **Tapia, Adams,** and **Percy Pierre** as a high barrier to increased minority participation in science and engineering. Adams cited standardized tests as "the gate-keepers to graduate school," and that graduate admissions officials are making their admissions decisions "like a man wearing suspenders and a belt, because he really has no faith in either one of them." Tapia argued that "the misuse of the standardized test is the underrepresented minori-ty's absolute worst enemy." Tapia has to work in a very creative way to deal with standardized tests like the SATs and GREs, and to get more minority talent into mainstream science in America and inside major universities. Tapia states that "we have to re-evaluate the evaluation criteria." **Lydia Villa-Komaroff** urged that the National Research Council simply eliminate standardized tests as a measure of their assess-ment of the quality of Research I institutions.

On the issue of race and the role of standardized tests - we must identify effective indicators and predictors of academic success among minorities. Pierre warned that "scientists have to stand up and say, statistically, if what we are trying to do is predict who is going to succeed academically and who is not, then this is not the way to do it." He further suggested that we need more longitudinal studies like the NCAA Proposi-tion 48 study to help us determine how to ascertain potential in science and technology, and to determine what indicators of academic potential are more valid than the standardized tests. Pierre described his research efforts in understanding standardized tests, particularly as they were used by the NCAA in establishing new criteria for eligibility of student-athletes. He further suggested that "we have to recognize the inadequa-cies of SATs as a predictor of undergraduate student graduation. Perhaps in determining who is going to do well in calculus I, SATs are a pretty good predictor. However, as a predictor of who is going to win the Nobel Prize, SATs are second-rate. I suggest rather than only focusing on re-evaluating the evaluation criteria, we should give some thought to more preparatory programs that bring disadvantaged youths up to the rigorous standards - empowering them like the Army does."

***

Closing Remarks

John Gibbons concluded the discussion by advocating that "a commitment to building a diversified science and technology community is to be valued in its own right. I don't think that requires quotas; neither does it require eliminating all of our science and engineering recruitment efforts for underrepresented minorities." Gibbons posed a challenge for us all to create a more supportive set of learning communities where government, university, industry and, most of all, individuals share the responsibility for attracting and nurturing diverse people into science and technology. **David Hamburg** stated that "this is not simply a meeting here. It is part of an ongoing process, and we will do everything in our power to see to it that the President continues to be oriented toward these issues...there is no substitute for Presidential leadership to stimulate the private sector, as well as the public sector. I think we have the opportunity to help the President do that. We can provide the leadership that could make a difference and to clearly make a diverse scientific and technological workforce part of a major thrust of the national agenda. It has got to be. I just want to emphasize that there is going to be a continuing effort flowing from this discussion."

APPENDIX A

Central Goals of
One America in the 21st Century:
The President's Initiative on Race

On June 14, 1997, President Clinton launched a year-long initiative, "One America in the 21st Century – The President's Initiative on Race," to examine the current state of race relations, look at the laws and policies that can help to ensure that we remain One America, and enlist individuals, communities, businesses and government at all levels to understand our differences while appreciating the values that unite us.

The President's Initiative on Race is an effort to move the country closer to a stronger, more just, and unified America, one that offers opportunity and fairness for all Americans. It is a chance for every citizen in our country to be a part of a great national conversation about America's racial diversity and about the strength it brings our nation. As Americans, our shared values unite us, but we can do more to be One America. The President is asking all Americans to join him in this effort which combines thoughtful study, constructive dialogue, and positive action to address the continuing challenge of how to live and work more productively as One America in the 21st century.

In announcing his Initiative, President Clinton stated, "Over the coming year, I want to lead the American people in a great and unprecedented conversation about race...What do I really hope we will achieve as a country? If we do nothing more than talk, it will be interesting, but it won't be enough. If we do nothing more than propose disconnected acts of policy it would be helpful, but it won't be enough. But if 10 years from now, people can look back and see that this year of honest dialogue and concerted action helped lift the heavy burden of race from our children's future, we will have given a precious gift to America."

The initiative will combine dialogue, study, and action. Its five central goals are:

1. To articulate the President's vision of racial reconciliation and a just, unified America.

2. To help educate the Nation about the facts surrounding the issue of race.

3. To promote a constructive dialogue, to confront and work through the difficult and controversial issues surrounding race.

4. To recruit and encourage leadership at all levels to help bridge racial divides.

5. To find, develop and implement solutions in critical areas such as education, economic opportunity, housing, health care, crime and the administration of justice – for individuals, communities, corporations and government at all levels.

APPENDIX B

Letters to Invited Panelists and Contributors

February 2, 1998

Panelist's Name
Street Address
City, State

Dear Panelist:

President Clinton has requested that I engage the scientific community in a dialogue on race as part of the President's Initiative on Race. In response, the White House Office of Science and Technology Policy (OSTP) and the Directorate for Education and Human Resources Programs of the American Association for the Advancement of Science (AAAS) are co-sponsoring a panel session at the AAAS 150th Anniversary Meeting entitled, *Meeting America's Needs for the Scientific and Technological Challenges of the Twenty-First Century*. The session will focus on (1) the need to diversify the science and technology community; and (2) how to deal with the challenges to targeted minority recruitment programs in science and engineering. I invite you to serve on this panel to share your advice and expertise on these issues.

The panel session will be held on Friday, February 13, 1998, from 2:00-4:00 p.m. at the Pennsylvania Convention Center in Philadelphia. I anticipate approximately 15-20 prominent scientists and industrial leaders to participate in the panel session that I will chair. Names of other invited panelists are attached. Also attached is a three-page document describing the broad outlines of the President's Initiative on Race and a summary of the Initiative's major events and accomplishments OSTP staff contacts are included on the attachment.

Earlier in the day, President Clinton will address the AAAS gathering to set forth his science and technology vision for the 21st century.

The President's address is scheduled for 12:30 p.m. at the Philadelphia Marriott Hotel (adjacent to the Convention Center). As a panelist, you are invited to attend. However, you must let us know by Friday, February 6, if you plan to do so.

I do hope you can join us, and I look forward to hearing from you.

Sincerely,

John H. Gibbons
Assistant to the President
for
Science and Technology

APPENDIX C

PANEL AGENDA:
OSTP-AAAS *ONE AMERICA* CONVERSATION

February 13, 1998
2:00-4:30 p.m.
Philadelphia Marriott Hotel

2:00 p.m.-   **Opening Remarks and Introduction of Panelist**
2:15 p.m.   John H. Gibbons
                Assistant to the President for Science and Technology,
                & Director of the Office of Science and Technology
                Policy

2:15 p.m.-   **Introduction and Opening Remarks of Moderator**
2:20 p.m.   David Hamburg
                President Emeritus, Carnegie Corporation of New York

2:15 p.m.-   **Panel Discussion**
4:15 p.m.   Speaker: Shirley Malcom
                American Association for the Advancement of Science

                Speaker: Cathleen Barton
                Semiconductor Industry Association

                Speaker: Julian M. Earls
                NASA Lewis Research Center

                Speaker: Carlos Castillo-Chavez
                Cornell University

                Speaker: James Comer
                Yale University

                Speaker: Richard Tapia
                Rice University

Speaker: Percy Pierre
Michigan State University

2:20 p.m.-   **Panel Discussion (cont)**
4:15 p.m.   Speaker: Samuel Massie
                US Naval Academy

Speaker: Lydia Villa-Komaroff
Northwestern University

Speaker: John Alderete
University of Texas Health Science Center

Speaker: Darleane Hoffman
University of California, Berkeley

Speaker: Howard G. Adams
Georgia Institute of Technology

Speaker: Satya N. Atluri
University of California, Los Angeles

Speaker: Fred Begay
*Assistant to the President of the Navajo Government
         for Science & Technology*

Speaker: O. LaVelle Bond
Procter & Gamble

Speaker: Herbert Z. Wong
Herbert Z. Wong & Associates

4:15 p.m.-   **Closing Remarks**
4:30 p.m.   John H. Gibbons
                Assistant to the President for Science & Technology

David Hamburg
President Emeritus, Carnegie Corporation of New York

Invited Contributors
George Campbell, Jr.

President and CEO, NACME, Inc.

William H. Gray, III
President, United Negro College Fund

Charles M. Vest
President, Massachusetts Institute of Technology

APPENDIX D

"One America Race Panel"
White House Office of Science and Technology Policy
(1999)

Dr. Howard G. Adams directs the National Institute for Mentoring at the Georgia Institute of Technology, and he is the former Director of the National Consortium for Graduate Degrees for Minorities in Engineering and Science. He is a recipient of the Presidential Award for Excellence in Science, Mathematics, and Engineering Mentoring.

Dr. John F. Alderete is the current President of the Society for the Advancement of Chicanos and Native Americans in Science (SACNAS).

Dr. Satya Atluri is an Institute Professor of Aerospace Engineering at the Georgia Institute of Technology and he directs the FAA Center for Aerospace Research & Education at UCLA.

Ms. Cathleen Barton is the Director of Education and Workforce Strategy for the Semiconductor Industry Association.

Dr. Fred Begay is the Assistant to the President of the Navajo Government for Science & Technology.

Mr. LaVelle Bond is Vice-President of Diversity for Procter & Gamble Worldwide.

Dr. Carlos Castillo-Chavez is a Professor of Biometrics at Cornell University, and is a recipient of the Presidential Award for Excellence in Science, Mathematics, and Engineering Mentoring, and a National Science Foundation Presidential Faculty Fellowship Award.

Dr. James Comer is the Maurice Falk Professor of Child Psychiatry at Yale University.

Dr. Julian M. Earls is Former Director of NASA Glenn Research Center at Lewis Field.

Dr. Darleane Hoffman is the Former Charter Director of the Glenn T. Seaborg Institute for Transactinium Science at the University of California, Berkeley. Dr. Hoffman is a 1997 recipient of the National Medal of Science for her distinguished work in Nuclear Chemistry.

Dr. Shirley Malcom heads the Directorate for Education & Human Resources Programs for AAAS and is a member of the President's Committee of Advisors on Science and Technology, and has recently rotated off the National Science Board.

Dr. Samuel Massie is Former Professor Emeritus of Chemistry at the US Naval Academy. Dr. Massie held an Honoree Chair of Excellence with the U.S. Department of Energy.

Dr. Percy Pierre is a Professor of Electrical Engineering at Michigan State University, Former President of Prairie View A&M, and Former Dean of Engineering at Howard University.

Dr. Richard Tapia directs the Department of Computational and Applied Mathematics at Rice University, and he is a member of the National Science Board. He is a recipient of the Presidential Award for Excellence in Science, Mathematics, and Engineering Mentoring.

Dr. Lydia Villa-Komaroff is Vice-President of Research at Northwestern University.

Dr. Herbert Z. Wong is a nationally recognized consultant of workforce diversity pioneer for Herbert Z. Wong & Associates.

Contributors

Dr. George Campbell, Jr. is the Former President and CEO of NACME, Inc., and a former member of the President's Information Technology Advisory Committee, Socioeconomic and Workforce Panel.

Dr. William H. Gray, III is Former President of the United Negro College Fund and Former Majority Whip of the United States House of Representatives.

Dr. Charles M. Vest is Former President of the Massachusetts Institute of Technology and is a former member of the President's Committee of Advisors on Science and Technology.

APPENDIX E

Position Papers of Invited Panelists

The Advancement of Science

by
Shirley M. Malcom

I am in science today because of two social movements, the "Space Race" and the Civil Rights Movement. The emphasis on the sciences was spurred by the launch of Sputnik by the then Soviet Union. At that time there was more attention to the science by the media, more opportunities for professional development in science and mathematics for our teachers, more resources directed to science and mathematics by government and a greater sense of "promise" in our segregated, resource-poor schools in Birmingham, Alabama.

Birmingham was also "ground zero" of the Civil Rights Movement. I still remember the turmoil of those times - the marches, the bombings, the attacks, living under curfew and martial law - and the resolve that we had as high school seniors in 1963 to go out and make something of ourselves.

Opportunities for African Americans had been so limited up until the mid-1960's and the possibilities so meager that even small cracks in the opportunity structure seemed like wide-open doors! Legislation and executive orders regarding equal opportunity and affirmative action represented promissory notes - now we would finally have an equal chance to achieve and succeed.

In Birmingham issues of differences were couched in terms of Black and White. Going to college took me to other parts of this vast country and expanded my view of the diversity of America. But, I was puzzled by the lack of diversity in my science courses - few women of any group, few African Americans, Latinos, or American Indians. In high school, every student in my chemistry, physics, and mathematics classes looked like me. In college, there were many occasions where I was "the only." At each succeeding degree level this became more the case.

Earlier this year my colleagues and I at AAAS received a grant from the Alfred P. Sloan Foundation to study both the current and recent changes in policies and practices of research universities with respect to graduate admission and graduate financial aid for underrepresented minorities pursuing science and engineering Ph.D.'s in research universities. The effect on professional school enrollment had been described for institutions in California and Texas, but less was known about effects on graduate programs. We surveyed 91 Research I universities - top recipients of federal R&D support that also enrolled and graduated significant numbers of underrepresented minority students at the graduate level. To augment our findings, site visits were made to a select group of institutions (ten), including those in states under legislative or judicial restrictions regarding the use of race as a factor in admissions or financial aid practices.

Our survey revealed that in many institutions no clear trend in enrollment yet existed for entering graduate students in science, mathematics and engineering. But the other clear signal was that many more institutions are experiencing decreases in minority graduate enrollment in science, mathematics, and engineering fields than are experiencing increases, and this among the top degree producing institutions. This finding does not bode well for the future production of minority Master's and Doctoral level scientists and engineers.

Perhaps more disturbing than the demographics is the climate that many students encounter: the isolation; the fact of being "the only one" in their departments or their classes; the search for a supportive learning community (as the social and the learning community and networks go hand-in-hand); the surprise of the faculty to see them; the different and lower expectations for their scholarly success. How much their stories sounded like my own story, 30 years out-of-phase!

But, we also found the isolated faculty member and department where diverse student groups were thriving. Our challenge is to create more supportive learning communities: where faculty share responsibility for attracting and nurturing a diverse group of students; where incentives are provided to do this; and, where the development of Latino, American Indian and African American scholars in science, mathematics and engineering fields is part of what it means to advance science and engineering.

Shirley M. Malcom is head of the Directorate of Education and Human Resources of the American Association for the Advancement of Science (AAAS).

Partnering for Workforce Development:
A Model for Increasing the Supply of Skilled Workers

by
Cathleen Barton

Key Points

Technology is the most important enabling industry in the world today. Technology represents about 50% of the growth of the US - including growth of the industry itself and cost savings from use of technology. It is the most important enabling industry in the world today. The shortage of technologically skilled workers is a fundamental threat to economic growth of the US; it hurts not only "high tech" companies, but the ability of the entire economy to grow by missing the productivity increases available with latest technology products.

Education for all, is a business, economic and workforce development imperative. The availability of quality education for all can no longer be viewed as a social and moral issue. It is imperative that education for all be recognized as crucial if individuals, communities, states and the country are to participate in economic growth, security, flexibility and prosperity. Women will continue to make up half of the workforce as we move into the 21st century; minorities are expected to increase to almost one third. Quality education for women and minorities is not a "nice thing" to talk about, or even do. It is a necessity. As we make preparation for the 21st century, we must all know that we will all be impacted if any one of us does not have a good education.

A global economy and marketplace. The best products for a global market will come from a diverse workforce representing the markets and communities in which we do business. In order to remain competitive in a rapidly changing global marketplace we must continue to attract and retain the most talented individuals in the world. The wide ranging experiences and perspectives of a varied population are crucial to continued success of our companies and our industry.

Awareness, ability and access. Prospective students must be aware of technology, the role it plays in our lives today, and the role it will continue to play. They must have the opportunity to be excited about it and challenged by it. They must have access to education and role models that will help them develop the abilities necessary and the resulting confidence to select post-secondary education in the areas of math, science, engineering and technology.

Teacher preparation. Teachers cannot teach what they do no know. They must be prepared, beginning with pre-service training, to effectively teach math and science. They must understand how to effectively use technology in their teaching. And they must have an understanding of the application of math, science, engineering and technology and opportunities in career choices. They should then be measured, recognized and rewarded in part on how well their students learn. Teachers are obviously role models as well. Increasing the number and retention of minority educators at all levels and all disciplines, and the number of women from middle school through post-secondary, engineering and technology related fields.

Post-secondary education increases outreach, changes learning environments. At the post-secondary level, increasing outreach programs for women and minorities continues to be a necessary and highly leveraged strategy in coordination with focused minority recruiting. Additionally, it is .incumbent on institutions to continue to hire women and minorities for faculty and staff positions. Finally, work must continue to create a environment where all students can be successful and achieve their fullest potential.

Partnering for workforce development: A model for increasing the supply of skilled workers. As the world becomes more technologically advanced, and few, if any industries, won't require some "high tech" workers, the U.S. semiconductor industry expects to be in competition with many of those industries for an insufficient supply of educated and skilled workers. Shortages for various types of technicians and engineers already exist in critical areas. Based on the input from its member companies, SEMATECH, a not-for-profit semiconductor manufacturing research and development consortia, agreed to sponsor a program that would address the projected shortage of skilled operators and manufacturing and equipment technicians. The program was initiated in June of 1996.

Program Goals

The main program objective is to increase the supply of skilled workers, operators and technicians for the semiconductor industry. The main goals are to increase the capacity to train skilled workers, focusing on the community and technical colleges as primary suppliers, and, to increase the enrollment and graduation rates for students in semiconductor manufacturing and electronics related programs.

The following is a more detailed explanation of the four project thrusts:

• **Capacity Expansion:** Increase the number and capacity of schools offering semiconductor manufacturing and electronics related programs at the post-secondary level.

• **Marketing Awareness:** Increase the number of students enrolled in semiconductor manufacturing and related programs by increasing the awareness of the semiconductor industry and semiconductor manufacturing as a career choice.

• **Alternative Sourcing:** Leverage the military discharge population and develop sources in addition to the traditional technician recruits. Alternative sources for re-training and re-careering, including a plan to identify "early-exiters" from four year colleges and universities.

• **K-12 School-to-Work:** Includes efforts at local, state and national levels to support strong mathematics, science and technology curriculums, and develop career and industry awareness opportunities for **teachers, parents, and students.**

In this initiative, in addition to establishing collaborative relationships at a local, state and national level between industry, education and government, a major component is creating awareness of and marketing technology careers in the semiconductor industry. We have implemented an "influence the influencers" strategy to create awareness using a combination of seminars through industry internships for students, teachers, counselors and parents.

Strong partnerships with community and technical colleges are generally consistent with corporate strategies to support the communities in which we work and do business. This education is broadly available and accessible to students and allows companies to develop and hire from a local pipeline.

In addition to a wide array of marketing materials developed for use by colleges and universities in recruiting students, many companies offer educational scholarships and frequently target them to help promote the entrance of women and minorities into high technology careers. Industry education partnerships at the colleges also focus on student retention. Some best practices currently utilized in this area include dedicated counselors, student study groups, co-op and internship opportunities and face-to-face and on-line mentoring activities. Work on "inclusive curriculum" and approaches are just beginning.

Results to Date

From June 1997 through September 1997, we:

- Increased by 50% the number of colleges offering semiconductor manufacturing programs from a total of 34 to 52.

- Increased enrollment in semiconductor manufacturing programs by 110% from a total of 2100 to over 4500.

Due to the continued demand and competition for technologically educated and skilled workers, the semiconductor industry has expanded its education and workforce focus. In addition to maintaining the technician focused Partnering for Workforce Development Program, the Semiconductor Industry Association currently sponsors initiatives in public policy, K-12 institutions, and four-year colleges and universities.

Cathleen Barton is director of Education and Workforce Strategy for the Semiconductor Industry Association.

Development Fund for Black Students in Science and Technology

by
Julian M. Earls

A group of black professionals in science and technology created an initiative to support black students pursuing degrees in technical career fields. The Development Fund for Black Students in Science and Technology (DFBSST) was established in 1983 by a group of concerned black professionals. The founders were: Dr. Julian M. Earls (NASA Lewis Research Center, Cleveland, OH) and Ms. Hattie Carwell (Department of Energy, Oakland, CA). In addition to these two, the initial members of the Board of Directors and their affiliations at that time were: Mr. James C. Jones, Jones Builders, Inc.; Mr. James Hicks, The Information Systems Corporation; Mr. Lee Browne, California Institute of Technology; Mr. Wayne Knox, Knox Consultants; and Mr. Knox Tull, Jackson and Tull Chartered Engineers. Members of the DFBSST believe it is essential for black professional scientists and technologists to provide financial assistance to talented black students who choose to enroll at Historically Black Colleges and Universities (HBCUs). The HBCUs traditionally have the highest retention and graduation rates for black students in technical disciplines. In addition, the greatest percentage of blacks earning doctoral degrees earned their undergraduate degrees from HBCUs. Therefore, our objective is to provide scholarships to black students pursuing technical undergraduate degrees full-time at these colleges and universities.

The DFBSST is a permanent endowment fund established in 1983. It is a 501(c)(3) tax-exempt non-profit organization incorporated in Washington, DC, and is solely devoted to providing financial support for targeted students. Deserving students are selected for support through the science and engineering departments at HBCUs. This support is increasingly important today when one considers the numerous challenges to scholarship programs targeted for people of color.

DFBSST members commit to minimum $1,000 annual contributions for life. Annual contributions less than $1,000 are accepted, but in contributor status rather than member status. This level of commitment is intended to demonstrate that black professionals are willing and able to help their own people and institutions. Black scientists and technologists of today owe those who preceded them a debt for the sacrifices and contributions, which opened doors to current opportunities. One way to repay those pioneers is to help open doors for students of today and the future. Therefore, members are required to make personal contributions towards the goal of an endowment of $1 million for scholarships.

Eligible students must be U.S. citizens who intend to pursue at least a 4-year course of study in science, engineering, or other courses in technology at an HBCU. There are 88 HBCUs, distributed throughout 22 states that offer undergraduate degrees in science and mathematics. Eight of these also offer degrees in engineering. Many of the other HBCUs are partners in dual-degree programs with other universities. Currently, the DFBSST solicits scholars at the eight engineering schools and seven other HBCUs with strong mathematics and science curricula.

Prior to 1989, the sole source of student candidates for scholarships was the National Merit Scholarship Corporation (NMSC) and its National Achievement Program for Outstanding Negro Students. Since 1989, to expand the student candidate pool, the DFBSST initiated an additional independent selection process, with the assistance of college professors of engineering and science. In both processes applicants are evaluated by criteria that include academic achievements, recommendations from teachers and counselors, self-expression as demonstrated by a written essay, and financial need. Of course the students must certify their intent to enroll at an HBCU or must currently be enrolled there and majoring in a field of science or technology. Financial need is used as a tie-breaker when applicants are considered to be equal based upon all other criteria.

Currently, the DFBSST is supporting 12 students. To date, the DFBSST has supported 48 students, 31 of whom have graduated. Students have continued through graduate school and have earned doctoral degrees.

Julian M. Earls is Former Director of NASA Glenn Research Center at Lewis Field.

Under-representation Perspectives from Academia

by
Carlos Castillo-Chavez

Youth media has proved to be a strong determinant of youth behavior and cultures. It has been used to foster racial stereotypes with movies like "Gone With the Wind," "West Side Story," and "White Men Can't Jump." Unfortunately, youth media, the most powerful approach for systemic change in education, has stimulated disinterest in science while keeping alive racial and gender stereotypes. No presidential program directed at diversifying the science and technological community will have a significant impact without the systematic cooperation and systematic participation of the media.

One of the saddest problems with this country is the state of public education, which is wildly inconsistent, even within schools. Clinton's program of "Getting Good Teachers Into Underserved Areas" is a step in the right direction, as good teachers are very badly needed. But why were they missing in the first place? At the heart of our educational differences is a public education system based on the "richness" of the local property tax base. Wealthy communities may spend as much as $14,000 per student while poor communities may not even spend a tenth of this amount. Economic inequities in the public school system foster sustained disadvantages among individuals and naturally affect people of color in a more dramatic way.

A Clinton plan without a systemic approach to the resolution of economic inequities in a property-tax driven public school system will have no substantial impact on the nation's efforts to diversify the scientific and technological communities. Furthermore, a lack of a systemic economic initiative in the way we fund public school systems combined with future racial demographic changes will create a two-class society that will rip our society apart. The president and congress must commit the time and resources needed to reform education. The president and congress must foster broad and constructive dialog on the impact of economic inequities on the scientific and economic fate of our nation. A leadership program that engages the country in the search for solutions for the nation's racial problems and its impact on the scientific enterprise is fundamental.

This leadership program, however, must continuously inspire all the nation's citizens to act on their own to create an America where equal opportunity is not measured by a SAT score, but by the ability of students to get a first-rate education from K-16 and beyond. Business, academic, and religious leaders must be critically involved in every *single* national and local effort that fosters diversity in the workplace. Leadership must come from the President, John Gibbons, and Congress.

Institutions of higher education have always been involved in the promotion of change at the national level from the time of sputnik. The national interest would be shortchanged without the significant and systematic participation of grade and high schools, technical schools, colleges, and universities in the development and implementation of the ideas, programs, and policies needed to help establish a diverse scientific and technological community. Systemic efforts such as those led by Luther Williams at the National Science Foundation (NSF) must not only continue but must also become an integral part of the portfolio of ideas and programs outlined on the President's Initiative on Race. Efforts to support individual scientists and mathematicians like Richard Tapia (Rice University) and Uri Treisman (University of Texas-Austin) who have developed and maintained research and educational programs that have dramatically improved the training and recruitment of underrepresented minorities into the sciences at the graduate and undergraduate level must be maintained.

Efforts to reward within the scientific community and within various university cultures -- the work that all faculty carries on the recruitment and training of minorities -- must be established in a systematic way. University administrators and faculty must be educated on the importance to the national interest of making sure that all our citizens have a clear opportunity to develop their scientific and technical talents.

At the moment most faculty, including most of our first-rate scientist and mentors, will not devote a sustained effort in the education, training, and recruitment of underrepresented minorities. Why can't our students have access to the best researchers and mentors in the world? Because these types of activities are not valued by individual departments, by university administrators, by federal agencies (such as NSF, NIH, NASA or DOD), and by private foundations (like the Sloan Foundation), which have systematically provided funds for summer research experiences and fellowships for a diverse population have been fundamental to the few successes that have increased diversity in our community of researchers. The sustained efforts of these federal agencies have resulted in the creation of high impact research programs for undergraduates that have already made a difference. The policies and initiatives of these federal agencies and the President's Initiative on Race must make it clear to university administrators and its faculty that support for these programs is fundamental to the nation's future.

Models of support and involvement by federal agencies must also be encouraged and rewarded. The systemic efforts of NSF's Luther Williams - MIE programs, for example - represent an extraordinary model of the role that NSF can play particularly when its director, Neal Lane, in this case, believes and supports the development of a diverse scientific and technological community. The individual effort of Jim Schatz at NASA provides a clear example of the impact that a single individual at a Federal agency may have on the nation's scientific infrastructure. Jim Schatz has devoted considerable amount of time in identifying and supporting programs at Berkeley, Cornell, and Rice that have become extremely successful. However, we must find ways of implementing successful individually driven programs, which often depend on successful and highly committed scientists into a broader community.

Programs like those supported by the Division of Mathematical Sciences at NSF and by the Division of Human Resources at NSF must have the resources to reach a significant number of students if they are going to make a difference nationwide. The creation of institutes that foster undergraduate research during the summer based on existing successful models is fundamental. Institutes must not only support research activities and mentor students, they must develop the expertise needed to implement them in diverse environments.

The establishment of institutes that bring the best faculty and the best institutional resources to students from all comers of the nation is urgently needed. To generate systemic change programs must be developed everywhere that include a significant number of the pool of eligible students. In order to generate systemic change, programs must not depend on the charisma or energy of extraordinary individuals who have taken upon themselves to address the issues of representation in academia. In other words, the essentials of programs that provide opportunities to a diverse community of students must be transferable to universities and local environments.

Faculty of all races will learn and implement successful models only if the value of recruiting, mentoring, and educating a diverse group of students is highly recognized and valued within each university and within each community. We are blessed with the most impressive scientific community in the world and yet most college students (of all races) have no significant opportunities to work with and be mentored by this invaluable resource.

Carlos Castillo-Chavez is former professor of Biometrics at Cornell University.

Diversifying the Science and Technology Community

by
Richard Tapia

In accordance with the session guidelines, I would like to make some comments directed at the following two concerns:

(1) The need to diversify the science and technology community; and

(2) How to deal with the challenges to targeted minority recruitment programs in science and engineering.

I begin with concern (1). The naive optimist states that the supply of scientists, engineers, and mathematicians, and the health of the science and engineering professions will be maintained by turning to the nation's underrepresented groups. However, the health of the science, engineering, and mathematics professions will be maintained by turning to foreigners. It is our history. We are quite good at importing solutions. We did it in World War II, when we had no scientists. We did in the 1960's when we had no mathematics faculty. We did it in the 1970's and 1980's when we had no graduate students.

While this has been the standard avenue for academia, today we see this approach practiced in a greatly increasing fashion by United States industry. The net effect is that underrepresented groups are becoming more and more of an underclass.

So one asks - why worry about under-representation, the health of the scientific professions is being maintained? The answer is quite fundamental and addresses a national crisis situation - "No first world nation can maintain its economic health when such a large part of its population is outside mainstream activity including all technological, scientific, and computational activity."

My message to the nation is that under-representation endangers the health of the nation and not the health of the various professions.

Improved representation will improve the health of the community and the nation.

Recently, I was discussing many of these issues with my friend and fellow National Science Board member, Eve Menger. She made a statement that states the case that I am attempting to present in a powerful and concise manner: "The greatest internal threat to this country is the formation of a permanent underclass. Good, public education is our best hope at prevention."

We have failed; and we will continue to fail if we don't increase our emphasis on the bottom-end of the pipeline. This means that we must deal effectively with society and the K-12 community. Today's minority youth picks up its values from the various segments of street and entertainment society, and not from parents and teachers.

I would like to now address concern (2). The recent Fifth Circuit Court of Appeals decision in the Hopwood suit against the University of Texas has had a devastating effect concerning the implementation of outreach programs directed at underrepresented minorities in the state of Texas. Anti-affirmative action backlash has been formally documented in California and Texas and will not remain restricted to these two states. It will take on a national posture. The Hopwood decision challenges us to learn how to continue outreach successes in a climate of anti-affirmative action sentiment and legal decisions. I view the position that we have been put in as a target of opportunity. I propose that the new flavor of affirmative action be what I have always interpreted it to be; an assessment and evaluation of the various evaluation criteria.

The use, perhaps better said, the misuse, of the standardized test at selective and even not so selective institutions is the underrepresented minority's worst enemy. My considered opinion is that this misuse is depriving the nation from tapping into a large part of its natural resources in terms of creativity and leadership. As such we are significantly retarding the process of moving along directions of change and reform that have been accepted as critical to maintaining our national health. For decades now, we have let the traditional beliefs of the ruling class dictate the policy for change and reform, and consequently we have ended up with an obvious lack of change and reform. It is imperative that we collect data, evaluate and assess, and use these findings as the impetus for change and reform. While we often allude to such studies, they are invariably incomplete, anecdotal, or non-rigorous. Hence, there can be no effective dissemination or buy-in on the part of our colleagues, administrators, and national educational policy makers.

I firmly believe that members of underrepresented groups, by the very nature of being a member of such a group, have learned skills and have developed sensitivities and understandings that would allow them to be more effective in various activities that we traditionally have valued and continue to value, and other activities that traditionally we have not valued, but have now realized that we must value.

For example, in research university environments we talk about the needs for nurturing, mentoring, more effective teaching, a better understanding of the whole student, and outreach to broader communities. Members of our so-called underrepresented groups are well positioned and prepared to contribute in these directions.

However, to a very large extent, these individuals do not have an opportunity to actualize or demonstrate this creativity and leadership skill because of traditional barriers. These barriers are not outright discrimination; no, they are much more subtle. They, on the surface, look like reasonable measurements of necessary prerequisites or skills. However, they are strongly biased towards the precocious attainment of various pieces of information and knowledge.

Potentials for success, creativity, the ability to guide and lead, the ability to adapt to a new environment and bring needed understanding from another environment, are not measured. This is too hard we do not know how to do this.

Moreover, our basic leadership is not totally unhappy with the process; since, after all, their careers were spawned by the process in place, so there must be some real good in this current traditional version.

While I am basically criticizing the use of standardized tests in undergraduate and graduate admission processes, it is a straightforward matter to extend my criticism to hiring policies, promotion policies, and selection procedures for prestigious fellowships, grants, and other professional rewards. We are in danger of locally restricting participation that would globally be of value to our national agenda. Local values and global values are usually at odds; indeed, often without being aware of this conflict.

It is time that we evaluate the evaluation criteria, its use, and its implementation.

Richard Tapia directs the Department of Computational and Applied Mathematics at Rice University, a member of the National Science Board, and a recipient of the President's Award for Excellence in Science, Mathematics, and Engineering Mentoring.

Race and the Misuse of Standardized Tests
in Predicting Academic Potential

by
Percy Pierre

General: The focus of the session is on (1) the need to diversify the science and technology community, and (2) how to deal with challenges to targeted minority recruitment programs in science and engineering. My comments will apply specifically to the engineering community and are colored by my twenty-five years of experience in creating programs to address issues of diversity in engineering. There are important differences between engineering and science that will affect how we address the problem of diversity. For example, in engineering, diversity concerns must include women. Nonetheless, I believe most of my comments are applicable to many of the sciences.

The Need to Diversify the S&T Community: There are compelling moral, ethical, commercial, political and legal reasons to diversity the S&T community. The reasons which seems to garner the broadest support is the need for skilled personnel to meet national economic and defense needs. In most fields of engineering, shortages occur on a regular basis. In the past, we have met that need with international students who have come here and stayed and often became citizens. Increasingly today, global technological companies are going off shore to buy engineering services from engineers often trained in the United States. Modern telecommunications and internet technology has made it possible for US engineers to team efficiently and effectively with engineers anywhere in the world.

More than one half of all engineering doctoral degrees granted in this country are granted to international students. With the engineering talent of the world available to US industry, it can be argued that there is no need to worry about US talents in these fields. I don't believe that it is good public policy to be too reliant on international sources when we have talent here to meet most of these needs. Minorities and women constitute two thirds of the US college-age population, yet only twenty percent of engineering doctorates granted to US citizens are granted to minorities and women. We need to do a better job. We particularly need more US minorities and women for faculty positions to help educate the engineers of the future.

Challenges to Targeted Minority Programs: Over the last twenty five years, I have helped create many educational programs designed to increase the number of minority engineers. While much remains to be done, these programs have been remarkably successful. Many believe that they have been the most successful set of minority programs in all of higher education. The recent challenges to affirmative action in higher education are a threat to many of these programs. While most of these programs can be improved and modified to accommodate current concerns, to abandon these programs would be a tremendous loss to the country.

The challenges to targeted minority programs are complex and the motives behind the challenges are many. Yet a recurring theme of the opposition is that scarce opportunities for education in selective programs should be granted on the basis of merit. Almost everyone agrees with that. It is the American way. Most also agree that those students most likely to excel in selective programs should be given preference. The problem is determining who is most likely to excel. A larger question is, can we have excellence with diversity?

This problem of achieving excellence with diversity is not peculiar to higher education in American society. The most competitive technological companies in the world make merit decisions everyday. Many do it while maintaining an increasingly diverse and global workforce. The US military has found a way to make merit decisions that keep it the best in the world while sustaining the most racially diverse leadership of any organization in American society. The biggest difference between the way these organizations determine merit and the way higher education determines merit is that higher education depends strongly on standardized test scores. Most American organizations employ more complex methods, which are both fairer and more effective.

The legal challenges to affirmative action in education are almost always posed in the following way. A majority student, who has been denied admission to a program, claims to be more qualified than a minority student, who was admitted, because the majority student has a higher test score than the minority student. This is a scientifically untenable claim. Standardized test scores are useful, but they should not be determining factor in deciding merit. The makers of most of these tests agree with this. The experts on test taking recognize that a two hour test is a very poor predictor of eventual success in a rigorous educational program. Success is usually dominated by other factors.

There is no better demonstration of the weakness of test scores as a predictor of success than the results of the NCAA longitudinal study of athletes from the early nineteen eighties. The NCAA Proposition 48 rule improperly used cutoff scores on standardized tests to help predict graduation of freshmen athletes. It was accurate in only 57 percent of the cases. This is little better than flipping a coin. Even worse, its failures were more frequent with white students than with black students. Forty percent of whites who made the score failed to graduate while only fifteen percent of the blacks who made the score failed to graduate. Fourteen percent of blacks who failed to make the score did graduate, while only six percent of whites who failed to make the score graduated. (see NCAA Report 93-08). Because of its inaccurate predictions, Proposition 48 unfairly favored white students. This fact is well known by experts, but they have failed to educate the general public, which still believes that using cutoff scores on standardized tests is fair.

Excellence with diversity is achievable, but it will require a more complex selection process than universities are accustomed to using or that the public understands. It will require the education of the public to the inadequacies of test scores in predicting educational success. Much can be learned from other American organizations that do achieve excellence with diversity. We need to assist universities in finding ways of doing this better.

The Office of Science and Technology Policy can help. First of all, it should encourage more longitudinal studies like the NCAA study to generate better predictors of success in science and engineering education. Secondly, it should encourage demonstration programs at all levels of education which would experiment with different means of selection that are designed to achieve both excellence and diversity. Race should be one of many variables in this process.

Race and Standardized Tests in Admission Criteria: Universities that have had to remove race as a consideration in admission have sought alternative admission criteria that would meet their particular needs. In California, state universities have removed race as a factor in admission but are uncomfortable with the resulting criteria which give inordinate weight to test scores. Nonetheless, the Governor of California insists that test scores continue to be the major factor in admissions. In Texas, the State Legislature has mandated that all students in the top ten percent of their high school graduating classes be admitted to state universities. This was an obvious circumvention of test scores which might pose a barrier to many students from high schools with low average test scores.

The University of Michigan continues to us race as one factor in admission even though this has been challenged in court and by many state legislators. Many private graduate and professional schools have discontinued the use of standardized tests scores in favor of better predictors of success in their programs. (Most still use race as a factor).

The central issue in the challenge to affirmative action programs in higher education is whether, and to what extent race and standardized tests scores should be used in admission criteria. This issue is often decided in a political context without solid scientific information on how best to use these factors to identify the best-qualified students.

There have been lots of statistical analyses done on standardized tests to determine their short term and long term reliability as predictors of academic success. These results all show that tests alone are weak predictors of ultimate educational success and that other factors dominate. In fact, the then President of the Educational Testing Service, which develops and publishes the SAT, strongly objected to how the NCAA planned to use SAT results in their athletic eligibility criteria. Yet athletic officials, politicians, lawyers and the public in general continue to put more reliance on test results than is merited by their statistical significance as predictors of educational success. Many do not trust educators to make admission decisions based on less quantifiable factors.

What is needed are studies that evaluate admission criteria, that will be credible to the experts as well as the general public. These studies should follow cohorts of students over long periods of time. They should test complex admission strategies involving many variables such as tests scores, grade point averages, letters of recommendations, personal interviews, curriculum evaluations, patterns of performance, various measures of student tenacity, race, and other factors. Most graduate programs are flexible and offer students different paths to the same result. A student may be more qualified for one path than another at the same institution. Admission criteria suitable to flexible programs should also be tested.

The design of these studies should call upon the best minds of statisticians, educators, and other experts in measuring human potential. A simple model would be to identify two cohorts of students who are admissible under two different sets of admission criteria and to follow them to determine their relative success in the same program or similar programs. This is what was done in the NCAA study. I'm sure there are better and more sophisticated ways of studying this issue. The kind of longitudinal study proposed here is seldom done because it is usually expensive to track students over long periods of time and the results are a long time in coming. That is why the federal government needs to be the supporter of this effort. These studies should be organized with broad user oversight in order to gain acceptance from educational policy makers and the public. The results can contribute to good public policy as well as better educational results.

The issue of the use of tests in admission is a broad educational issue. It is different at the undergraduate level and the graduate level. At the graduate level, each discipline usually sets its own admissions criteria. Here, I am especially interested in science and engineering. A few well designed longitudinal studies of admission criteria at the graduate level in science and engineering would make a significant contribution to the broader issues.

I would hope that the White House Office of Science and Technology Policy (OSTP) would initiate appropriate programs in this area.

References

NCAA Research Report, 93-08, National Collegiate Athletic Association, Overland Park, Kansas, July 1994.

New Directions in Assessment for Higher Education: Fairness, Access, Multiculturalism, and Equity, The GRE, FAME Report Series (Vol. 1), Graduate Record Examinations, Educational Testing Service, Princeton, New Jersey.

At the time of publication of this document, a new decision in the courts recently surfaced regarding race and the role of standardized tests. On March 8, 1999, a Federal District Court in Philadelphia discarded the National Collegiate Athletic Association's (N.C.A.A.'s) minimum test score requirement for athletic scholarships, throwing into disarray a long-accepted criterion for establishing eligibility for student-athletes. The court held that the N.C.A.A.'s S.A.T. and A.C.T. minimum test score requirement, as outlined in Proposition 16, "has an unjustified disparate impact against African-Americans." In reaching its ruling, the court cited the N.C.A.A.'s own research showing that the practice harmed African-American students' chances of being declared academically eligible, and that the organization's goal of improving graduation rates -- the reason it instituted Proposition 16 in the first place -- could be achieved by other available methods. The court agreed with a Washington-based trial lawyers advocacy group representing the plantiffs in the class-action suit that the minimum test requirement was a violation of the 1964 Civil Rights Act, and that the N.C.A.A. could be sued under that act, which prohibits race discrimination by educational institutions that receive Federal funds.

Percy Pierre is Former Vice President for Graduate Studies and Professor of Electrical Engineering at Michigan State University.

Achieving a Diverse Science and Technology Community

by
Samuel Massie

The need to diversify the science and technology community and methods of meeting the challenges to targeted minority recruitment programs in science and engineering are two to the problems which the engineering profession, through its AMIE Committee, has been discussing for many years. The AMIE Committee, a coalition of representative and engineering professionals from Fortune 250 companies, nine Historically Black Colleges and Universities (HBCUs and one Hispanic Association of Colleges and Universities (HACU) was created in 1992 to provide a proactive approach to cultivating diversity in engineering as an essential business strategy along with other related objectives. The name AMIE is taken from the first letters of the purpose of the organization, Advancing Minorities Interest in Engineering.

Because the objectives in both the AMIE studies and the AAAS panel are similar, I wish to make some comments, which apply to both areas, based on my position as Honoree Chair, but I state firmly that some of my recommendations may have other original authorship.

Discussion I. The Need To Diversify The Science and Technological Communities

- Projected demographic populations indicate that an increasing proportion of our available employment force will be minorities and women, individuals, who normally are not included in our projected work population. Attracting, educating, recruiting and promoting diverse, high-performance teams of unique individuals - each with valuable talents and strengths - is critical to maintaining a competitive advantage as we approach the 21st century. As a nation, we can no longer afford to overlook this valuable national resource.

- Diversification does not always mean different races, sexes or cultures - it may mean different and new ideas and successful adaptation of them. Many of the jobs now needed were unknown ten years ago. It is likely that ten years from now our job needs will be for jobs that are presently unknown.

- The information age is greatly changing our social and technical needs, just as the industrial age created new employment needs in the 20th century.

- Present day technical education must contain two related factors - a combination of technical depth and social breadth.

Discussion II. Methods of Meeting The Challenges To Targeted Minority Recruitment Programs in Science and Engineering

- Before we can consider methods it may be wise to consider some of the challenges. The May 1997 ACME Research Letter points out four special problems: (1) The number of minority freshmen enrolling in engineering programs and science in the United States has dropped in the past three years. (2) Scholarship funds are becoming less available to minority students, especially with recent judicial statements. (3) Changes in financial aid packages are forcing minorities to incur increased debt, and with limited promises of upward mobility payment of debt is perceived as uncertain. (4) Newly emerging firms are not diversifying their staffs and certain medium to large size corporations which traditionally offered the highest starting salaries, have downsized their operations.

- Diversity is often confused with affirmative action. Prejudice is often defined as being down on something you are not up on.

- The promised federal assistance to community colleges and technical education may be very useful!

The Late Samuel P. Massie is Professor Emeritus of Chemistry at the US Naval Academy. Recently, he was honored with the Dr. Samuel P. Massie Professorships of Engineering Excellence in the Environmental Disciplines at nine Historically Black Colleges and Universities (HBCUs) and one Hispanic Serving Institution (HSI), sponsored by the US Department of Energy's Office of Environmental Management.

On the Retention of Underrepresented Minorities in Science

by
Lydia Villa-Komaroff

To increase the number of underrepresented minorities in science, there are three areas that must be addressed: getting children through high school with the expectation of going to college, retaining in science those who enter college with an interest, and increased entry into graduate and professional training. I will comment only on two of these areas.

Early Childhood. In early childhood, we need to maintain the interest that every child has in science. Particularly with underrepresented individuals, we must allow our children to envision a future beyond that of their immediate circumstances. I have never met a child who was not intensely interested in the way the world works. However, most of the time, this interest is lost by junior high. In addition, children seem to conclude very early that there are some professions that are not open to them. Our children need to see images of people they can identify with as scientists. We must give them a sense of possibility and the belief that they belong in the mainstream world.

There are a number of experiments that have been tried and are successful, but they must be expanded. One such program is "Mothers and Daughters," a program based in El Paso, Texas and conceived of and led by Josephina Tinajero, a professor in the school of education at the University of Texas, El Paso. There are four keys to this program: first, getting the kids and parents involved when the children are very young, second, heavily involving at least one parent, third, providing experiences that provide the child with the feeling that they are both capable of and entitled to a college education, and finally, providing adult role models.

Professor Tinajero, herself a child of a Texas barrio, looked at the high levels of teen pregnancy, the low high school graduation rates, and the essentially nonexistent rates of college enrollment that occurred in children who grew up in the poor neighborhoods of El Paso. In these families, English was spoken poorly if at all, there was no family history of higher education, and very low expectations.

Professor Tinajero decided that any intervention needed to occur before the children entered adolescence, when peer pressure and hormonal changes make outside influence difficult, so she decided to focus on fifth graders. In acknowledgment of the importance of family, she also decided that the mothers must also be involved in any program. So she choose families, often recent immigrants, where no family member had gone to college. The parents spoke little or no English. The daughters and mothers then entered a one-year program where they would visit the University of Texas campus several times, at least once for an overnight visit in a dorm. They met and talked to students. They had visits and were able to ask questions of Mexican American women in a variety of professions - policewomen, lawyers, judges, accountants, scientists, writers, airline pilots, for example. At the end of the year, there is a ceremony where the mothers and daughters make pledges to each other. The daughters pledge to finish high school, to not get pregnant until after marriage, etc. The mothers pledge to help the daughters find a place to do homework, to support their daughter's ambitions. The pledges are written by the participants on heavy paper provided by the program. These often are framed and placed prominently in the homes. The program can maintain only minor contact with the students after the one year. The remarkable thing is the results - the program, when I heard about it in 1995, had 300 girls go through the program and get to age 18 - all had completed high school, only 3 had become pregnant, a large fraction had enrolled in college and were acting as "big sisters" for new 5th graders entering the program. At least equally remarkable, the mothers also were affected by the experience, as a number of them returned to school and entered the workforce.

Retention. The only other area I will address is the need to expand the number of minority students who graduate from college and who enter graduate programs. There is substantial documentation that students who graduate with degrees in science, math or engineering entered college with that interest. Furthermore, many students who began college with an interest in science, math or engineering leave the fields before graduation. Essentially no students are recruited into science majors in college.

One of the barriers to minority participation is the heavy reliance on test scores for entry into undergraduate and graduate programs. This use continues even though it has been documented for some time that, above a certain threshold, these scores are not good predictors of success. I believe that there are two factors that have made it difficult to loosen the reliance of schools on these scores. The first is that these tests are universally available and appear to be a single measure that can be used to rank a large number of students. Although it is clear that test scores are not good predictors of success, it is not clear what factors are more predictive. A better use of these scores would be to set a threshold score. All students above that threshold would then be judged on other criteria like class rank, high school or college grades, references, written essays, etc. This approach has been used with good success at Rice.

The second factor is that the average score of admitted students is one of the measures used to rank undergraduate institutions and graduate programs. I believe that institutions would be more willing to use other criteria if their own standings were not so heavily dependent on them. I propose that the National Research Council stop using average scores as a measure of graduate programs and that US News and World Report do likewise.

Lydia Villa-Komaroff is Former Vice-President of Research at Northwestern University.

Absence of Minorities From Research Fields
Will Result in Grave Consequences in U.S.

by
John F. Alderete

Must we continue to remind ourselves about the under-representation in the entire science enterprise of our country of African Americans, Hispanic-surnamed Americans (Mexican Americans and Puerto Ricans), and American Indians? The complexity of this issue that America is faced with the under-representation of minorities in science makes it one of the most challenging facing our country today. What can America expect if it does not correct this exclusion of a large proportion of its citizens in all research fields?

Consider the demographics of Hispanic citizens in the United States, keeping in mind the other two prominent underrepresented groups, African Americans and American Indians. Today according to US Census data, one of every five of the nation's eighth-graders is Hispanic, and most have at-risk attributes (from a single-parent home, low parental education, limited English proficiency, low family income, no role models because siblings are dropouts, and spending more than three hours each day alone). Only 53 percent of these Hispanic students will finish high school. According to data from the Hispanic Association of Colleges and Universities, only one in nine of those who finish high school will attend a four-year university, and it will take that student 12 years to finish four, mostly because of poverty!

It is appreciated that minority professionals make a difference in the health of our country's minorities. Yet, out of 15,365 M.D.'s who graduated in 1992, only 632 (4 percent) were Hispanic-surnamed. Consider that Hispanics will soon surpass African Americans as the largest U.S. minority group and will approach 100 million in number in the U.S. in less than 50 years. Things must change - and soon - or our country may experience unforeseen events, such as the L.A. riots and decreased support from minority taxpayers for research universities and government agencies, with its minority groups on the fringes, un-empowered, disenfranchised, and undereducated.

The equation that accounts for the absence of minorities in research is exceedingly complex, as evidenced by examining variables that impact negatively on minorities:

1. The latest annual report on world hunger by *Bread of the World*, a citizen's lobby group on hunger issues, estimates that 4 million American children under the age of 12 are hungry and 9.6 million more are at risk of hunger. The National Center for Children in Poverty (Columbia School of Public Health Report, 1994) indicates that one of every three minority children in the U.S. is poor, and there are pockets throughout our country, such as South Texas, where one-half of minority children less than five years of age are poor. Poverty, woven into the fabric of minorities' lives, contributes to multiple factors that place them at risk. Poverty contributes to overall poor living conditions, sub-optimal health, lack of health insurance, poor nutrition, inaccessibility of adequate health care, and lack of transportation to health care facilities. Chronic hunger, accompanied with weight loss, headaches, fatigue and loss of concentration, undermines the ability of poor children to learn.

2. Minorities are disproportionately affected by disease because of where they live. A 1993 report from the UCLA Center for Occupational and Environmental Health shows that, in Los Angeles, 50 percent of all Latinos live in industrial areas with the most polluted air. Blood lead levels of minority children exceed normal levels. Children of farm workers have exceedingly higher rates of cancer and birth defects than the national average, for example, presumably owing to pesticide exposure. Eighty-five percent of the new HIV/AIDS cases in U.S. cities occur in African American and Latina women (Dateline: NIAID, September 1996). Smoking, alcohol, substance abuse, cancer, sexually transmitted diseases, and infectious diseases are, in many cases, higher among minorities (CDC Morbidity and Mortality Weekly Report, 1995).

3. Among the criteria for admission into research universities is an over-reliance on standardized test scores. Moreover, few research-intensive institutions have targeted efforts to recruit minorities, and there appears to be a lack of mentoring of minority students. Many institutions recruit in foreign countries but never set foot in minority-serving institutions. Articles in two recent issues of *The Chronicle of Higher Education* (B.J. Fraser, October 31, 1997, page A58; G. Masien, November 28, 1997, page A48) described the increased recruitment in Latin America by U.S. colleges and the need for universities to develop new strategies to compete for students from Asia, respectively.

4. There is a negative impact on the numbers of minorities applying for and accepted to all professional schools caused by accepted anti-affirmative action rhetoric that is inflamed. Further, recent court rulings, such as the Hopwood Decision of the 5th Circuit Court of Appeals, and the anti-minority sentiments that are being expressed openly in some universities have the net effect of further distancing graduate programs from qualified minorities.

Despite these variables, there are data to show unprecedented numbers of minorities and women graduating from B.S. degree-granting institutions with interests in pursuing higher education. This notwithstanding, many of these deserving American students will not be admitted to professional programs if they target recruitment efforts toward foreign students and rely on standardized tests as a major criterion for admissions.

If our nation is concerned about the fact that so many of its minority citizens are undereducated and underutilized in higher education, and if it understands the concomitant costs associated with the deterioration of our youth's and, therefore, nation's social health (such as child abuse, teenage suicide, drug abuse, and poverty (Institute of Innovation in Social Policy Report, Fordham Graduate Center, 1996), then what systemic changes are needed? We need new paradigms of education and must expect new and innovative programs to deal with the losses that occur throughout the educational pipeline. The 50% success rates for graduation from high school of Hispanics is unconscionably low and must change. Majority schools must embrace affirmative action-like methods to recruit and admit minorities form underprivileged backgrounds. Majority, publicly supported research institutions must develop or augment existing remedial programs and internship work-study initiatives. We must urge all minority programs to make sure that support for minority students is, in fact, directed to those whom it is intended, and not to resident immigrants or non-citizens. Universities must require faculty to participate in community science activities and view the promotion of education and mentoring of minorities as meritorious as conducting extramurally funded research. We must keep in mind that a significant proportion of our country's citizens are minorities. They work hard and pay taxes and therefore deserve to have their children educated to the highest levels.

John F. Alderete is a professor of microbiology at the University of Texas Health Science Center in San Antonio and president of the Society for the Advancement of Chicano and Native American Scientists. E-mail: alderete@uthsesa.edu [Portions of this commentary will appear in *The Scientist* February issue dedicated to minority issues in the sciences.]

Why America is Still "A Nation at Risk"
Fifteen Years Later?

by
Darleane Hoffman

As long ago as August, 1981, then Secretary of Education T.H. Bell created the National Commission on Excellence in Education to examine the quality of education in the US and issue a report within 18 months of its first meeting. The impetus for this was concern about the "widespread public perception that something is serious remiss in our educational system." The Commission was chaired by David P. Gardner and included public and private school and college educators, presidents of colleges and universities, science professors and researchers, and even one Nobel Laureate in Chemistry, Professor Glenn T. Seaborg. The Commission's report entitled, "A Nation at Risk: The Imperative for Educational Reform" was issued in April, 1983 and its findings in assessing the quality of teaching and learning in our Nation's public and private schools, colleges, and universities, and in comparing American schools and colleges with those of other advanced nations and the declining performance of the students and graduates shocked the nation. However, the Commission also tried to define the problems to be faced and overcome in order to successfully pursue the course of excellence in education. The Commission was directed to pay particular attention to teenage youth which they did by focusing largely on high schools, but selective attention was given to the formative years in elementary schools, to higher education, and to vocational and technical programs.

A few quotations from their report, issued nearly 15 years ago, are still relevant today.

The first sentence is: "Our nation is at risk."....

"America's position in the world may once have been reasonably secure with only a few exceptionally well-trained men and women. It is no longer."....

"All, regardless of race or class or economic status, are entitled to a fair chance and to the tools for developing their individual powers of mind and spirit to the utmost. This promise means that all children by virtue of their own efforts, competently guided, can hope to attain the mature and informed judgment needed to secure gainful employment, and to manage their own lives, thereby serving not only their own interest but also the progress of society itself."....

"Learning is the indispensable investment required for success in the *information age* we are entering."....

"For our country to function, citizens must be able to reach some common understandings on complex issues, often on short notice and on the basic of conflicting or incomplete evidence. Education helps form these common understandings."...

The report quotes John Slaughter, a former Director of NSF, who warned of "a growing chasm between a small scientific and technological elite and a citizenry ill-informed, indeed uninformed, on issues with a science component."

"Knowledge of the humanities...must be harnessed to science and technology if the latter are to remain creative and humane just as the humanities need to be informed by science and technology if they are to remain relevant to the human condition."

After considering this though provoking document and examining what I see of the situation in science today and for the 21st century, I have come to the conclusion that if we are to attract, foster, and utilize the talents of a more representative cross section of our population, in both race and gender, we must start before high school and college. Although we have been rather successful with mentoring programs at the high school and college levels, these are primarily for students who are already interested and wish to pursue a career in science or mathematics.

However, if we really wish to draw a larger fraction of minorities (and women) into careers in science, we need to start education in science at the kindergarten level and carry it on through the elementary school level and into high school. We must recognize, however, that not all children will want to (or should) pursue careers in science, any more than all should become professional musicians or newspaper reporters, although many may well wish to acquire knowledge about the skills involved. However, all should have an understanding of the scientific method and some basic knowledge and literacy in science and mathematics, if they are to become the informed electorate who must make choices concerning energy use and production, environmental concerns, health concerns, nutrition issues, transportation and safety, the risks, benefits and economics of their choices, etc. Additionally, non-specialists need to know where to find information and how to assess it.

In order to accomplish the diversity of goals ranging from attracting children into a focused scientific career to educating the informed lay person required to function in our technological society, may require a different approach to teaching of science than we have taken in the past. This is particularly important for children who do not grow up in a "scientific" environment. It does not need to involve sacrificing the content of the science that is taught, but it may require considerable upgrading of the quantity and quality of the science education of the teachers involved and implementing of standards.

Such programs are currently being developed. For example, the PRIME Science Group (College of Chemistry) UC Berkeley has developed a program beginning with grade 6 and going through high school that is designed for different learning styles and interests. It has complete teachers aides, guides, texts, and training and developmental workshops. Activities are written at more than one level of complexity based on the students learning style, relevant prior experiences, interest in that topic, etc. these are designed around topics that all would agree are of general interest, e.g., "Keeping Health" in which illness and the role of microbes in infectious diseases are examined, and "The Earth is Restless" in which data on earthquake intensities are analyzed. Students particularly interested in a given topic can explore it in more depth. There is provision for homework and after school activities as well. The high school program would reinforce and expand on previous themes in life and earth sciences, chemistry, physics, and physical science and is aligned with the National

Science Education Standards. The approach lends itself well to "team and modular teaching" and the use of modern computer technology; however, this does not substitute for actual "lab" experiments.

It appears that the above techniques could also be extended to kindergarten through fifth grade levels. Furthermore, there are many simple concepts and even experiments that could be introduced even at the pre-school level such as magnetism, density and buoyancy, and you can all probably think of your own favorites.

Darleane Hoffman is the Charter Director of the Glenn T. Seaborg Institute for Transactinium Science at the University of California, Berkeley. She is also a 1997 recipient of the National Medal of Science for her distinguished work in Nuclear Chemistry.

Addressing the Issue of Underrepresentation of Minority Groups in Graduate Engineering and Science Education

by
Howard G. Adams

Graduate schools through credentialing and certification serve as the 'springboard' to positions of leadership, power and authority (Adams, 1993). For underrepresented minority groups (minority groups) the springboard (graduate and professional schools) has and continues to be out-of-order or closed. Institution report attempts to serve minority groups with the hollow statement "we tried to diversify our programs without success." And because of their lack of success, minority groups continue to be underrepresented among engineering and science (E/S) doctorate degree graduates (less than 3% in engineering and less than 6% in the hard sciences; NCR data).

Addressing the continuing underrepresentation of minority groups in E/S graduate education is a major challenge facing the Nation as we prepare for the twenty-first century. We can not hope to ensure equity of opportunity for minority groups as professors, researchers, managers, leaders, policy makers, etc., if we fail to provide access and choice of E/S graduate education at the highest levels.

Key questions that will need to be considered as the Nation formulates its response are:

1. How to improve current methods for identifying and recruiting qualified, capable minority groups to full-time E/S graduate education (Adams, 1997)?

> **NOTE:** Qualified minority applicants exist, but because of lack of information and/or encouragement, they elect not to pursue E/S graduate studies.

2. How best to structure the admission process to include criteria for assessing who is worthy of admission to and capable of suc-

ceeding in E/S graduate education (Adams, 1997)?

> **NOTE:** Standardized test scores are relevant measures, but it is well known they fail to tell the whole story of qualification. Other measures such as motivation, previous work experiences, and personal achievement can also contribute to admission information.

3. How to ensure that adequate financial aid is available to support minority group E/S graduate students (Adams, 1993)?

> **NOTE:** Currently, approximately 55% of all faculty controlled E/S graduate student research aid goes to foreign students with less than 5% going to minority students. Foreign students are 10-1 more likely to be funded by their department than are minority students (NSF data).

4. How to create graduate program environments that are conducive to developing the full academic potential of minority group E/S graduate students (Adams and Conley, 1986; Adams, 1993)?

> **NOTE:** Because of poor advising and mentoring, minority students are outsiders to their departments; have limited opportunities to serve as TAs/RAs; are denied access to departmental resources; are really "graduate school orphans" that no one wishes to adopt.

Providing greater access and choice for minority groups to E/S graduate education is an achievable goal. However, it will take more than "good intentions." What is needed is a commitment to sound, constructive policies and action steps for implementation. We can continue with the status quo or initial bold and new approaches. Just remember, "If we do what we've always done; we will get what we've always gotten - continued under-representation of minority groups in E/S graduate programs.

References

Adams, H.G. and Conley, M.M., "Minority Participation in Graduate Education: An Action Plan," The Report of the National Forum on the Status of Minority Participation in Graduate Education, Washington, DC, 1986.

Adams, H.G., "Focusing on the Campus Milieu: A Guide for Enhancing the Graduate School Climate," The National Consortium for Graduate Degrees for Minorities in Engineering and Science, Inc., Notre Dame, IN, 1993.

Adams, H.G., "Recruiting Graduate Students: Implementing the Key "R's" of Graduate Education," The National Consortium for Graduate Degrees for Minorities in Engineering and Science, Inc., Notre Dame, IN, 1997.

Howard G. Adams is Founding Director of the GEM National Institute on Mentoring (NIM) headquartered at the Georgia Institute of Technology, College of Engineering, Atlanta, GA.

Personal Reflections on Race and Achieving S&T Diversity

by
Satya N. Atluri

Since this panel is convened to engage the scientific community in a dialogue on race, I express my views on this subject in specific. Also, since I am not a social scientist or a philosopher, I think it is best to bare my soul of my personal experiences as they relate to race as factually as possible, and draw some conclusions from them - in the traditions of engineering and the physical sciences that I am trained in.

I grew up in India - where race and ethnicity were not as important factors as caste and colors of the skin (the various tones of brown). Two hundred years of colonial rule have brought the common man in India to accept "white" as being the "superior" color of the skin. In the India of my youth, 'diversity' meant diversity of language, diversity of caste and of religion. In the 'Indian Institute of Science' where I was schooled, my graduating class of 30 did not consist of more than 2 students from any linguistic group - and it was my only real experience in 'diversity.' We couldn't communicate in any language other than English, which was not the mother-tongue of any of us.

It is with this background that I came to M.I.T. as a student in 1966 - and instantly underwent intellectual and personal experiences that changed my life forever - for the better, I though at that time. My doctoral advisor at M.I.T. was born in China; the professor of mathematics, with whom I developed an intellectual kinship, was a Jewish person who left Germany for America in 1936; my roommate in the dormitory was a very nice Englishman; the students with whom I had the greatest intellectual rapport were from EVERYWHERE on the globe; the first woman with I developed an intellectual and emotional relationship was from Chicago; - all in all M.I.T. was a kaleidoscope of racial, cultural and ethnic diversity, at least from a student's point of view. These were the best years and had profoundly affected my life. I left M.I.T. in 1971, thinking that someday I will return there, to contribute to making it even better.

In the years 1976-1988, there were several discussions of my returning to M.I.T. - but never in a concrete way. In December 1989, I was surprised to receive a phone call from the Head of the Aeronautics and Astronautics department, inviting me to come to M.I.T. as the Jerome Clarke Hunsaker Visiting Professor during 1990-91. This, I knew to be a fairly prestigious position; and I was overjoyed at this opportunity to re-experience all that was good in my life until that point. Instead, what really resulted was a nightmare - the year 1990-91 turned out to be the most humiliating period of my life. I was made to feel most unwelcome by a majority of the faculty.

Indeed, I was told that the prospect of my potentially coming to M.I.T. on a permanent basis was unacceptable to a number of the faculty there. This sentiment of the faculty resulted in my being treated in a most insulting way. A couple of years later, an untenured faculty member (a Caucasian) who left M.I.T. told me that, prior to my arrival at M.I.T. as a visitor, the sentiment was expressed at one of their faculty meetings that "Asian faculty belong at state universities and not a M.I.T." I think that in a faculty of nearly 40 members, there are no Asians in that department now; and I believe there is only one member who belongs to a racial minority at M.I.T. and other science and technology universities of comparable stature and "faculty-driven" universities.

The notion of diversity - is then left to the discretion of the faculty. In intellectually competitive and charged environments such as M.I.T., this is analogous to a group of little children, all of the same race, having the sole power to decide who else can enter the playground. It is in this sense no laws on "equal opportunity" have any real meaning. It is imperative that our political leaders and intellectual leaders constantly remind everyone of the necessity for diversity as in all our institutions, including our most elite ones.

I greatly applaud President Charles Vest of M.I.T. to have done this in his recent annual addresses to his faculty colleagues at M.I.T. I hope that more and more of our intellectual leaders will similarly embrace the need for diversity.

In the last few decades, Americans have come to appreciate the global nature of our economy and of our industries. In the next few decades, human resources, and intellectual capital, will be the most important assets of any nation. More and more, intellectual capital will have no national boundaries. If one visits the emerging elite centers of science and technology in Asia - in Hong Kong, in Korea, in Singapore, and in Japan - one notices the enormous push to bring the world's best talent to these universities.

Our country, which has been a magnet for this talent in this century, should do whatever to make it still attractive for the world's best and brightest to come here. Rather than indulging in "tokenism" of diversity at all our institutions, we should put our resources in sustaining and creating a few "magnet institutions" of higher learning where the best and the brightest, irrespective of their race, gender, and ethnicity, can thrive in a true atmosphere of "diversity." Along with that, we should provide ample access to underrepresented minorities to positions of leadership at these institutions - even at "faculty-driven" elite universities. This would produce the much needed future leadership in science and technology in America, with racial diversity.

Having visited a large number of countries during my academic life, there is no question in my mind, even from a racial tolerance point of view, ours is the greatest country. Since ours is a unique nation in the human history, it is incumbent on all of us who have benefited from the institutions of our country, to make these institutions perfect - and I believe "diversity" is the key to this perfection, for the good of everyone, but most importantly for the majority race!

I would like to close on a very personal note. In March 1991, during my visit to M.I.T., I was diagnosed with the end-stage renal disease - kidney failure.

In November 1992, I was fortunate to receive a cadaver kidney transplant in Atlanta. I am alive today due to the generosity of a family who made a caring decision even in the time of their grief - I am not aware what the race of this family is - nor, I think, they know my race. Acts of grace and kindness do not involve race - and one only hopes that these acts place every minute at every institution in our country.

Satya N. Atluri is director of the Center for Aerospace Research and Education in the School of Engineering and Applied Science at UCLA.

Beyond Conflict or Compromise

by
The Late Fred Begay

Background

Today leaders of the Navajo Government are addressing problems and challenges faced by developing countries in other parts of the world. The Navajo leadership is determined that the Navajo Government's resources including appropriations from the U.S. Congress, be effectively directed to the needs of the Navajo people, and to a course of development consistent with Navajo values and goals.

An important objective for the Navajo Government is to provide its people with the essentials of a decent life, including adequate nutrition, housing, employment, earnings, education, health care, consumer goods, and public services.

The Navajo development projects should focus on one basic target, the fuller and better use of available natural resources and human talents and energies. The goal of development is to improve the total environment and to better the quality of life in all possible aspects.

The Navajo has experienced over the past century two traditional ways of resolving extreme differences with the U.S. Congress. **Conflict -** which is always the worst. **Compromise** - which is usually the best. But suppose a situation arises in which conflict threatens unacceptable risks, and compromise unacceptable losses? What then? Or rather, what now?

A third way must be found to effectively resolve the complex structure of Navajo human and natural development problems.

State of the Navajo Government

Approximately 220,000 Navajos reside on a 17.5 million acre Reservation located in the States of Utah, Arizona, and New Mexico. The

Navajo Reservation was established by the U.S. Congress in June 1868 Treaty. No civil rights were incorporated into the 1868 Treaty. Over the course of the past century, the Navajo Government has filed civil rights law suits against the U.S. Congress, and the U.S. Supreme Court has generally ruled in favor of the Navajo.

Selected 1990 socio-economic indicators show that (1) the personal income per capita is $4,106, (2) unemployment varies from 50-80%, (3) average school years by adults is 10.5, and (4) percent homes with plumbing is 13%.

There are 160 K-12 Navajos schools with approximately 60,000 students managed by the States of Utah, Arizona, and New Mexico, Federal and Navajo Governments, and Church Institutions.

Developing the Navajo Human and Natural Environment

Today, approximately one billion dollars is needed to support human and natural environment programs. General categories of development projects include:

(1) *Developing the Human Environment* - rural and urban communities, industry and cash crop agriculture, education and training, capital requirements, population, social progress, health, and
(2) *Developing the Natural Environment* - animal life, plant life, water and soil, weather and air, minerals.

Seaborg Hall of Science

The most important measure of success for the Navajo Government is its ability to make improvements in the lives of Navajos. Few enterprises touch the lives of as many people as do those concerned with education and training. High-quality education and training benefit the individual whose knowledge and skills are upgraded, the business seeking a competitive edge, and the Navajo Government in achieving overall productivity and increasing competitiveness in the global marketplace. It is essential that all Navajos have access to the education and training they need and that the teaching and learning enterprise itself become a high-performance activity.

In recognition of the importance of education and training...and of the necessity of bringing to bear the talent and coordinated resources of all Navajo education institutions to create productive and successful education and training enterprises that support the lifelong learning needs of all Navajos...the Seaborg Hall of Science (SHS) was established in January 1998. The SHS is an independent non-profit institution dedicated to the encouragement, support and patronage of learning through research and scholarship across a wide range of fields in the natural and social sciences. The SHS will provide a vital resource to support the collaborative work of the Navajo, State, and Federal Government toward directing expert attention to the solution of Navajo development problems.

The SHS will advise and assist the Navajo Government in coordinating and increasing the effectiveness and productivity of Navajo efforts in education and training. Its purpose is to:

(1) Coordinate Federal, State, Navajo, and Church education and training programs;
(2) Promote the use of technology to enhance lifetime learning; and
(3) Promote excellence in K-12 science and mathematics education.

The SHS Vision: To coordinate and focus Federal, State, Navajo, and Church efforts in education and training so that they become a powerful force in helping Navajos meet the challenges of the 21st century.

The SHS will develop its Strategic-Implementation Plan to guide Navajo efforts in education and training in accord with principles established by the National Education Goals as delineated in the "Goals 2000: Educate America Act," Public Law 103-227 (March 31, 1994), the National Science Education Standards, National Research Council Project (1995), and "Curriculum and Evaluation Standards for School Mathematics," National Council of Teachers of Mathematics, 1989. The Strategic-Implementation Plan will provide a framework for making cross-institutional policy, programmatic, and budgetary decisions, and for assessing the impact of these decisions.

The Late Fred Begay is the former Assistant for Science and Technology to the President of the Navajo Government.

Meeting America's Needs for the Scientific & Technical Challenges of the Twenty-First Century: Procter & Gamble's Goals and Perspective

by
The Late O. LaVelle Bond

Importance of Technology to Procter & Gamble (P&G)

We at P&G place great importance on technology, on finding the best people - to continue our fine tradition of providing innovative, superior products to the world's consumers. There are important technical challenges at the heartbeat of P&G's business. The innovations that result from these technical challenges are what consumers know us for: our brands.

We sell more than 300 brands in over 140 countries. We have 17 major technical centers worldwide; on-the-ground operations in nearly 70 countries, and about 100,000 employees. Our net sales are more than 35 billion dollars, and we are one of the largest consumer goods companies in the world.

We are a Company of brands, but the success of our brands is tied directly to the excellence of our technical disciplines. You've probably been led to believe that we are a marketing company: in fact, we are an innovation company. Product innovation has been the cornerstone of our success in the past, and it is our primary strategy for success in the future.

Statement of Purpose

Our Statement of Purpose says: "We will provide products of superior quality and value that improve the lives of the world's consumers." We've been successful because we have lived our Statement of Purpose every day since our Company was founded in 1837. Over the past 160 years, Procter & Gamble has established a record as a pioneer in developing new technologies and innovations in consumer products categories that affect the health and well-being of consumers worldwide.

In some cases, we're not just competing in these categories; we're literally creating them. For example, (1) Crisco, the first all-vegetable shortening, (2) Tide, the first heavy-duty synthetic laundry detergent, (3) Crest, the first flouride toothpaste proven to effectively prevent cavities, (4) Head & Shoulders, the first pleasant-to-use shampoo effective against dandruff, (5) Bounce, the first dryer-added fabric softener, (6) We virtually invented the disposable diaper category with Pampers, (7) We were the first with a shampoo and conditioner in one, introduced in Pert Plus, and (8) We were the first to offer people a zero-calorie, zero-fat additive with Olean.

National Medal of Technology

Procter & Gamble was recently rewarded for innovations like these with the National Medal of Technology. This is the highest honor given in the United States for achievement in technology.

In presenting the award, President Clinton recognized P&G for creating, developing and applying advanced technologies to consumer products that have helped improve the quality of life for billions of consumers worldwide.

Key Diversity Related Awards

1994 - U.S. Secretary of Labor presented P&G its highest award - Opportunity 2000 - for instituting comprehensive work force strategies to ensure equal employment opportunities.

1995 - Leadership Conference on Civil Rights awarded retired Chairman and Chief Executive Officer Edwin L. Artzt, its Private Sector Leadership Award in recognition of the Company's efforts and cooperation in bringing together the business and civil rights communities to work for equal rights for all Americans.

1994 & 1995 - P&G was one of the 50 U.S. companies recognized by VISTA magazine for its leadership role in offering career opportunities to Hispanic women.

Additional Perspectives

Technical innovation is fundamental to our business, and the men and women who work in our technical functions are the heart of P&G's innovative capacity.

A global survey of 300 chief technical officers recently identified P&G as one of the world's top 10 organizations for technical excellence.

We back up our commitment to innovation with an investment of nearly $1.5 billion a year in research and development.

In 1996, P&G field for 19,600 patents worldwide. That's a 12% increase over 1995, and a 63% increase over 1993. We have more Ph.D. scientists than are on the MIT and Berkeley faculties combined.

Recruiting Results for Technical Discipline

During the past 5 years, 62% of our total management and professional hires were from technical disciplines. About 29% were minorities.

Development and Support of a Diverse Workplace

The intent at Procter & Gamble (P&G) is to develop all employees to their full potential. To achieve this goal, P&G has human resource systems in place that support individual development, and the company regularly reviews these systems to make sure they work well for everyone, including women and the various population groups that make up their minority workforce. For example, the number of women and minorities at the director level and above, has doubled over the past five years.

Ongoing support systems in place include:

- Career discussions, performance appraisals, assignment plans, transfer and promotion plans.
- Grass roots/informal network support groups which have ex-

isted at different sites for a number of years.

• Mentoring to provide informal support and guidance in addition to coaching and training provided by each employee's direct manager.

In-depth diversity reviews are conducted regularly with each organization head. P&G has established five-year goals and year by year plans for every individual judged capable of assuming greater responsibility at some point in the future.

P&G builds into their culture the values, behaviors and norms that create advantage from differences. P&G has become increasingly diverse, and as the company has grown around the world, it has become even more important that P&G fully utilize this individuality. The company workplace environment encourages collaboration which brings the employee's different talents and experiences together to produce better ideas and superior services and products.

Measurement of Progress in Achieving a Diverse Workforce

P&G monitors its progress in achieving a diverse workforce in a number of ways including:

• P&G has a company officer heading the company's global diversity organization.
• Diversity goals are established for each organization and are included as one of the key measurements in achieving overall business results.
• Employee performance reviews are based on a list of "what counts" factors which include a employee's ability to respect and work effectively with diverse people. This gives individuals direct feedback concerning their personal progress. For employees who have others reporting to them, their performance review also include assessments of their ability to develop people, including women and minorities.
• Regular employee surveys are conducted that collect feedback about the health of the company's culture and environment over-

all, as well as how P&G is doing company-wide in building diversity into all organizations and levels. The surveys have shown there is increasing awareness of the importance of diversity.

Diversity Training for Employees

P&G has a number of diversity training programs available. Many are conducted at the corporate level, but most are conducted by the company's business units and manufacturing plants. A combination of in-house and external resources is used to conduct the training, with some of the programs developed by the company and other developed by outside consultants. Diversity also is a key concept interwoven throughout many of the training courses P&G offers, including P&G College.

The Late O. LaVelle Bond is the Former Vice President of Diversity for Procter & Gamble Worldwide.

**Workforce Diversity in the Science and Technology Community:
Key Strategies and Future Directions**

by
The Late Herbert Z. Wong

Introduction

We are, today, at the "cross roads of change" in workforce diversity. While on the one hand, we are seeing ever increasing and promising efforts in the focus on workplace diversity, we are at the same time seeing the divisive and erosive efforts of State legislations, hate crimes, and reported discrimination and harassment in the workplace.

For example, with respect to positive diversity efforts, there were: (a) more than a dozen of national conferences in the past year attended by 200 to 600 participants each, (b) newer and better tools and materials for diversity programs and related activities, and (c) diversity initiatives in over 50% of America's businesses with over 100 employees. On the negative side, we are seeing the divisive and erosive effects of (a) legislations like Proposition 209, in California and in other states, (b) hate crimes and reported discrimination and harassment in the workplaces and marketplaces, and (c) less confidence by the American public that our racial and multicultural relationships are getting better (with even less confidence by ethnic and racial "minorities").

At no time in our history, do we feel greater "anxiety and fear" over the possible danger and barriers to diversity nor at once greater "hope and promise" of possible opportunity and gains from our workplace diversity initiatives in our business, educational, governmental, and community organizations.

While we can each take important individual roles, ultimately and imperatively the challenge must be met by our leadership and organized efforts within our "institutions." To have impact, our diversity management and leadership strategies and efforts need to occur within our educational institutions, our business and corporate structures, our governmental agencies and branches, and our community (church, synagogue, temple, and other voluntary organizations). Success rests upon our organized, interactional efforts -- strategically (by planned and precise actions), systematically (by well-orchestrated communication), and collectively (through inclusion of all).

I would like to focus my remarks on what we have found effective in workforce diversity in organizations. We will apply these key strategies and practices to achieve solutions for diversifying the science and technology community and for dealing with targeted minority recruitment programs in science and engineering.

Background

In 1987, the term "workforce diversity" was populated by the Hudson Institute's Study, *Workforce 2000, Work and Workers for the 21st Century,* and the term "valuing diversity" became a call for action throughout corporate America as a result of the three-part videotape series *Valuing Diversity* by Lewis Griggs. At the same time, "diversity training" by Herb Wong and other colleagues was the response by corporate America to the call for action.

By 1991, the Conference Board's survey conducted by Louis Harris & Associates, found that 79% of the businesses and corporations surveyed had implemented activities or had future plans for implementing a diversity initiative (primarily through diversity training). In that same year, at the First Annual National Diversity Conference, over 350 participants discussed the variety of activities and major components of their organizations' diversity strategies and initiatives. In the 1993 the Society for Human Resource Management (SHRM) survey, 61% of the respondents indicated that it was a "business necessity" to conduct diversity training. In 1995, *Training Journal,* in its annual Industry Report, found that over 50% of U.S. organizations with 100 or more employees were conducting some form of diversity training and/or diversity program. Clearly, diversity initiatives are seen as central to the success of many organizations today.

While cross-cultural communication, cultural differences, and multicultural training and education dates back to as early as the 1960's in higher education, the emergence for the understanding of "workforce diversity" (in the Hudson Institute's sense) and the notion of "diversity initiatives" parallel the time frame and activities found for corporate America.

Now, with more than a decade of experience in diversity initiatives, what "key strategies" and "best practices" have emerged? Our review of the existing reports and literature resulted in over 100 cited "best practices." Based upon our review and our experience, we have condensed and highlighted the following overall summary of "key strategies" and "future directions" for diversity initiatives as applicable to the science and technology community:

There is a Current Drive for Unity in Organizations

- Diversity is defined from an inclusive perspective.

  It is defined as "all the various characteristics that make one individual the same or different from another." This definition is used to prevent employees from feeling excluded in efforts to enhance diversity within the organization. The overriding principle is inclusion and involvement -- "no one stands outside of diversity." Some organizations have attempted to focus on the ways we are the "same" in an attempt to link similarities to unity within the organization.

- The perspective of diversity as a unifying factor is central for the organization (through focusing on dimensions of diversity (e.g., age, education, gender, race, etc.) that are similar among people as well as different.

  Employees want to have "healthy" and "value-added" interactions and relations within the workplace and want to "heal" work-relationship tensions and conflicts. An organizational philosophy embracing pluralism as a definition of diversity is adopted. "Targeted minority recruitment" is viewed as one of the tools to develop the pluralism desired in the organization. It shows diversity as not a program with a start and finish, but an ongoing process that will change the makeup of the organization.

Eliminate Misunderstandings About Diversity

- The organization provides clear delineation and definitions for its workforce diversity efforts.

   There is an alignment of diversity to the organization's overall and specific strategic and business goals, and activities are well coordinated and clearly articulated. The context of (e.g., educational and business organizations), process used in (e.g., targeted minority recruitment), and goals to be achieved (e.g., diversify the science and technology community) for diversity initiatives are well understood by the organization.

- Diversity initiatives are clearly linked to other strategic initiatives of the organization.

   The function, structure of, and context for workplace diversity to other organizational and business initiatives and strategic plans are well understood.

- All diversity activities start with and maintain senior leadership support throughout the organization (and throughout all phases of the diversity initiatives).

   Senior management commitment, presence, and support is visible; senior leaders serve as "Ambassadors for Diversity." Diversity is approached and supported from the "top down" and "tailored to the organization." While it is not as restrictive as, *one size fits only one*; it certainly is almost never, *one size fits all.* "Off-the-shelf" diversity programs, without any adaptation to the organization, have resulted in limited value and success.

Some Key Strategies and Future Directions for Workforce Diversity

• We must focus on not only changing behavior, but "changing the heart."

> Diversity initiatives and programs are not just about changing the actions and behaviors of individuals (a very important first step). It must move beyond "just tolerating differences" and "being politically correct." We must understand, believe in, and value the similarities and differences that we bring to our organization. Certainly, the dialogue on race as part of the President's Initiative on Race moves us in that direction.

• The organization derives its diversity strategies and diversity activities from information based upon current, valid, and reliable needs assessment and/or organizational culture audit findings.

> Workplace assessment information allows diversity programming to utilize the "strengths" of the organization to overcome the "barriers" to inclusiveness. It capitalizes on the "opportunities" for change while vigilant in terms of the decisive issues that could bring unexpected conflict within the organization. The organization listens to the "voice of its members, employees, and customers."

• The organization needs to discover, understand, and intervene in the organizational factors related to workplace diversity (e.g., career development, supervisory leadership, performance evaluation, rewards and recognition, mentoring, etc.).

> Effective diversity strategic initiatives rarely are just activ-

ities associated with "diversity" programs. An effective organizational diversity strategy must go beyond programs, processes, and activities identified solely with the diversity program initiative.

• The organization develops a culture of inclusion and fairness, and communicates and implements organizational values, behavioral norms, and performance standards toward the elimination of harassment and discriminatory behaviors.

It recognizes and rewards those who value, promote, and facilitate workplace diversity, and establishes an organizational culture and climate that is antagonistic to those who harass and discriminate.

• Eager participation and visible commitment of senior management, as well as program leadership, and active managers in diversity programming initiatives are critical.

Top management puts diversity in the forefront of the organizational posture (on "center stage") and communicates its personal commitment regularly and forcefully. The organization has a diversity "center" to provide key points of contact and/or foci for all diversity-related activities.

• The organization recognizes and articulates the "extremely compelling" needs and reasons (from the unique perspective of that organization) for diversity programming initiatives.

The services and business connections need to be made. The organization links and aligns its diversity goals to corporate strategy.

• The organization resolves problems that, for some employees, often feel quite major -- in resolving these, recruitment and retention are supported.

The organization has a well-established process for reporting and for problem resolution on diversity issues. Managerial training on diversity conflict resolution and diversity performance coaching have been favorable.

• Guiding principles for diversity are provided to management and employees.

These are simple communications that provides a rationale for supporting diversity. Model diversity letters for executives and managers to send to their subordinates reflecting the organization's posture and initiatives are provided annually. This expedites communication, making it easier, consistent and more likely to actually get done. The organization communicates effectively to create buy-in; then it communicates more.

• The organization understands its "readiness to change," and its leaders understand their role in managing the organizational change process.

Senior management understands the risks and barriers; develops contingency plans; and has a continuous process for diversity programming improvement.

• The organization develops benchmarks and "best practices" comparisons.

Most have developed "internal" best practices (within organization best practices). There are a few "industry" best practices comparisons (to include the same or different organizational functions and to include competing and/or non-competing organizations). We are currently developing some exciting national best practices information and comparative norms for U.S. organizations, and we will then have a way to go for global standards and "world-class" best practices for diversity program initiatives.

• The organization has a process of accountability and monitoring of its diversity programming initiatives.

> Some organizations provide specific monetary bonuses and other rewards linked to accomplishment of stated diversity outcomes with established metrics for measuring success.

• People's expectations about diversity training and related program initiatives are well managed.

> Realistic outcomes and anticipated events and changes need to be carefully managed. Ongoing discussion and communication relative to people's expectations are vital activities.

• Consultants (internal as well as external) provide valuable resources and objective perspectives.

> Central to the ability of the consultant to help the organization is the assessment of the "goodness-of-fit" of the Consultant to the needs of the organization. The greater the clarity of the organization's expectations, needs, barriers, and opportunities relative to its diversity initiative, the easier it is for the organization to successfully select its Consultant and for the Consultant to be successful in implementing the diversity initiatives of the organization.

• Executive coaching, diversity mentoring, and/or management learning ("lab" or training group) are complementary resources for middle to upper managers to strengthen their people management skills as an augmentation to diversity training.

• Clearly stated and understood outcomes for the diversity initiative for individual employees, business units and groups, the organization as a whole, etc. not only help the organization to manage expectations but allow accurate methods and metrics to evaluate diversity training results and organizational impact.

Conclusion

I have briefly summarized for you some of the "key strategies" and "future directions" for diversity initiatives based upon the research literature and out practical experience. I am pleased that this is a field that is challenging, rewarding, and contributing to our people, our organizations, and our communities. I hope that my remarks will be useful toward diversifying the science and technology community.

The Late Herbert Wong is a nationally recognized pioneer of fostering diversity and inclusion in the workplace. He co-founded the National Diversity Conference in 1991, and co-authored "Multicultural Law Enforcement: Strategies for Peacekeeping in Our Diverse Society." He was nationally acclaimed as a workforce diversity consultant for Herbert Z. Wong & Associates. He was the founding Executive Director of the Richmond Area Multicultural Services (RAMS) Comprehensive Community Mental Health Center, which provided mental health services in 17 languages in addition to English.

APPENDIX F

Position Papers of Invited Contributors

Engineering and Affirmative Action: Crisis in the Making

by

George Campbell Jr.

Engineering: Past Exclusion, Recent Progress, Uncertain Future

The early 1970's, the aftermath of the Civil Rights Movement, held forth great promise for America's minority populations. Jim Crow laws had been, at least formally, abolished. The Civil Rights and Voting Rights Acts had been passed and institutionalized. The Federal Government was vigorously enforcing Equal Employment Opportunity regulations. Nowhere was the promise greater than in engineering. African Americans, Latinos and American Indians were then virtually invisible in the profession, comprising barely one percent of the work force. But the nation's consciousness had been raised about the unhealthy consequences of the historical exclusion of minorities which had led to the severe under-representation that existed.

Engineering, the nation's largest profession, was the nucleus of economic development and wealth creation. More than half of the CEOs in Fortune 500 companies were engineers. Management at all levels was dominated by individuals with engineering backgrounds. Clearly, if minorities were going to achieve upward mobility in the corporate sector, we had to produce more minority engineers. Just as clear, the engineering profession was being deprived of an enormous wealth of talent, given that a major segment of the population was essentially left out. Economic, social, political and legal pressures for change were prevalent.

In this context, the NACME organization was created to lead a national private sector effort to create broad access to the engineering profession. Corporate, government and academic leaders were eager to lend their enthusiastic support to this effort, and a number of other organizations, programs and outreach efforts were created. Progress over the past quarter century has been no less than astounding. From less than one percent of the engineering work force, minorities have grown to almost six percent. Universities produce more than 6,000 minority BS graduates in engineering annually, up from several hundred at the end of the Civil Rights era. Since 1980, NACME alone has invested more than $100 million in its mission, and ten percent of all minority graduates since then were supported by NACME scholarships.

The story, however, does not yet have a happy ending. The 6,422 African American, Latino and American Indian bachelor's degrees in engineering in 1997 comprised only ten percent of the total graduates. The 197 minority doctorates graduated in 1997 amount to only 2.8 percent of the total – this from population groups that comprise 28.5 percent of the college-age population. And impeded by the double bind of ethnic and gender bias, minority women, at fifteen percent of the college-age population, constitute an important group to consider separately. They make up only 1.1 percent of the engineering work force and, in 1997 received only 2.8 percent of the bachelor's degrees and 0.6 percent of the doctorates in engineering.

More discouraging than the graduation numbers is the current climate of resistance, the nationwide assault on equal opportunity efforts, the social, political and legal pressures that diametrically oppose those of the 1970's. This climate not only threatens further progress but jeopardizes the gains already achieved. The data are clear. For the four year period from 1992 to 1996, minority enrollment declined a whopping 10.4 percent. There were 1,574 fewer minority freshmen in the entering engineering class of 1996 than there were in 1992. Minority enrollments, which had bucked the declining trend for non-minorities during the previous decade, now reinforce rather than mitigate that trend (8.9 percent decline from 1992 to 1996).

Growing Demand for Engineers

While the talent pool of African Americans, Latinos and American Indians continues to be underutilized in engineering and while enrollment declines precipitously among all ethnic groups, the demand for engineers and other technically trained people is growing explosively. During the past four years, actual engineering employment increased from 1,717,000 to 2,051,000, a growth of almost 20 percent. The current unemployment rate for engineers stands at a near-record low of 1.5 percent, less than a third of the unemployment rate for the work force as a whole (4.9 percent). According to a recent survey by the Information Technology Association of America, there are at this moment almost 200,000 unfilled jobs in computer and information technology. The average starting salary for 1997 engineering graduates with just a bachelor's degree approached $40,000, which was 40 percent higher than that for business administration graduates, 75 percent higher than that for liberal arts graduates, and rivaled the starting salaries for 1997 law school graduates.

To meet our work force needs, the United States has come to rely heavily on foreign-born engineers. In 1995, more than 65,000 of them immigrated to this country. That's as many engineers as American universities produced that year. Foreign-born engineers comprised more than 40 percent of the graduate school enrollment, received almost half of the doctorates awarded and held more than 60 percent of the post-doctoral R&D positions in 1995. Forty percent of the doctoral engineers resident in the United States are foreign born, as 30 percent of the engineering faculty at American universities.

While promoting xenophobia or recommending immigration curbs would be counterproductive, it's essential for us to understand the dynamics of our technical work force. In the oil crisis of 1973, we saw what can happen when we depend too heavily on foreign sources for critical commodities, and in today's world, human capital is our most valuable resource. Recent changes in the increasingly global labor market suggest that the United States is, in many ways, becoming more vulnerable and less favorably positioned to attract well-trained engineering talent. Competitor nations are expanding their investments in research and development relative to GNP, as we contract ours. State-of-the-art R&D laboratories that rival the best American labs are being constructed at universities, government and industrial facilities overseas, attracting graduates who, in the past, would have been more inclined to remain here after their education. Some foreign governments have structured financial incentives to attract top expatriate scientists and engineers back to their homelands. Clearly, continuing over-reliance on foreign engineers while neglecting the development of intellectual talent from segments of our own population is very risky. It's also unnecessary.

Shifting labor costs on an international scale for science and engineering professionals present a great opportunity for our scientific enterprise, but crystallize another emerging risk. Once topping the charts in labor costs for scientists and engineers, the United States is now positioned very competitively. This should reduce one of the incentives that led American companies to export an increasing number of R&D jobs in recent years. On the other hand, if we do not produce the necessary work force to meet the growing demand, American companies will be forced, not only to export more R&D jobs, but often to pay higher labor costs while doing it.

Assault on Affirmative Action

While our failure to develop the talent pool of minorities for the engineering profession threatens America's economic well-being and corrupts our moral fabric, the anti-affirmative action movement, that is well-organized, well-funded, politically connected and effectively articulated, is sweeping the nation. The fictional rhetoric, reflected in much of the press coverage, incorrectly equates affirmative action with preferential treatment on the basis of race, an egregious misrepresentation of the concept and original intent of affirmative action policy. Anti-affirmative action resolutions call for an end to racial preferences, when, in fact, federal regulations have always prohibited such preferences. Affirmative action does acknowledge, however, that we are a society that is not yet color-blind, a society with a history of deeply rooted exclusionary practices, a society that demands proactive policies to ensure equal opportunity and to eliminate both conscious and inadvertent discrimination. In making admissions or hiring decisions, both subjective and objective metrics that have inherent biases are commonly used. Without explicitly recognizing those biases in interpreting the metrics, equal opportunity and social justice will not be possible.

The 1996 decision in *Texas v. Hopwood* by the United States Fifth Circuit Court of Appeals prohibits the use of race or national origin as a factor in public university admissions. This case was not merely about the personal harm Cheryl Hopwood suffered as a result of her rejection from the University of Texas Law School in favor of minority students with lower test scores. More *non-minority* students than minority students with lower test scores than Ms. Hopwood were admitted to the law school. This case was part of a measured, coordinated national effort on the part of opponents of affirmative action to attack the policy. The Center for Individual Rights which represented Ms. Hopwood recently filed similar suits against the University of Washington on behalf of Katuria Smith who was denied admission to the university's law school in 1994 and against the University of Michigan on behalf of Jennifer Gratz and Patrick Hamacher. The latter case significantly raises the stakes. It challenges the undergraduate admissions policy for the first time and also, in seeking monetary damages from the president and former president of the university, imposes the threat of personal liability, thereby increasing the pressure on individual college administrators to back away from affirmative action.

The impact of the Hopwood decision is evident from the significant decline in first year enrollment of minorities in the University of Texas Law School from 1996 to 1997. African American enrollment dropped from 31 to 4, Latinos from 48 to 26 and American Indians from five to three. The representation of minorities in 1996, with a first year class size of 502, was already unreasonably low considering that minorities comprise 44 percent of the college-age population in Texas. At the undergraduate level minority freshman enrollment at the University of Texas declined from 17.4 percent to 15.1 percent, a thirteen percent drop in the proportion of minorities, at a public institution, supported by tax dollars, in a state where everyone pays taxes at the same rate (Texas has no graduated income tax). At Rice University, one of the selective private schools in Texas, the proportion of minorities in the freshman class dropped a third, from 18 percent to 12 percent. In a state where minorities will soon constitute a majority, these are, at best, foreboding trends.

In California, another state that will soon have a "majority minority" population, affirmative action policies have also been successfully contested. The California Board of Regents passed a resolution banning affirmative action in university hiring and admissions in 1995 and, in 1996, California voters passed Proposition 209. A voter initiative to eliminate affirmative action, the measure was recently upheld by the courts. The impact on university enrollments has been similar to that in Texas.

Some analysts have conjectured that, although the proportion of minority students declined at the prestigious institutions, other institutions within the affected states may be increasing enrollment. Even if this turns out to be true, we must ask ourselves whether or not that is a desirable outcome for our society. Currently, minority engineering graduates are distributed among the universities categorized by selectivity in roughly the same proportions as non-minorities. That is, the most selective colleges produce about 10 percent of the minority engineering graduates and 10 percent of the non-minority graduates. The next tier produces about 25 percent of both minority and non-minority graduates, etc. Although students at the undergraduate level can arguably acquire a quality education at a second tier school that will not differ substantially from that at a top tier school, we may assume that career paths on average are at least loosely correlated with where individuals attend undergraduate school.

Perhaps the greatest danger resulting from recent court decisions is that they drive us further and further toward the imprudent practice of relying on rigid numerical results of standardized tests in determining the qualifications of students for admission to the university. We already know that, for minority students, standardized tests do not yield a good measure of academic capabilities. The Scholastic Assessment Test (SAT), for example, is more reflective of students' parental income than of their academic potential and more highly correlated with parental education than students' eventual success. It's no surprise, then, that minority students do not do as well on SATs as their peers. The significance and relevance of the test outcomes also vary by race and ethnicity. While the correlation between test scores and eventual academic performance is low for all groups, it's virtually nonexistent for minority students.

Raw test scores, without taking family background into account, measure past opportunity rather than future academic potential, and are clearly biased along ethnic lines. Given that they're not predictive of academic success, and that the asserted goal of the university admissions process is to select students most likely to succeed, the tests are non-functional. These facts are widely known; however, beyond the legal environment, other factors compel universities to cling to the use of SATs and to seek students with high scores. The average student SAT score, for example, is a key parameter used in university rankings.

At NACME, we have a great deal of experience and considerable success using more authentic assessments in the selection process for students in our programs. These involve measurable attributes which are much more highly correlated with academic and career success: creativity, problem-solving skills, approach to unfamiliar problems, motivation, determination, interest, commitment, discipline, reliability. We have students in our programs who started college with SAT scores as far as 600 points below the average of their peers but who are graduating with honors from the most prestigious engineering colleges in the country.

It's encouraging that, in both Texas and California, universities are diligently working within the new legal framework to find solutions that will allow them to maintain their goal of building a diverse student body and their commitment to the rich intellectual environment it produces. A University of California task force has recommended that the university system eliminate the use of SAT scores from admissions considerations. In Texas, the legislature voted to require state universities to admit all applying students who graduate in the top ten percent of their high school class regardless of SAT scores.

At least 16 states have substantial anti-affirmative action activities underway, including litigation, voter initiatives, pending legislation or executive actions. Borrowing from effective strategies of the mid-century

Civil Rights Movement, anti-affirmative action political groups have carefully researched specific candidates for court challenges, catalogued individual judges' records and scoured the nation to identify optimum legal venues for anti-affirmative action litigation. Not surprisingly, their efforts are focused in those regions where there are large concentrations of minorities and where there is already limited access. This means that successful anti-affirmative action outcomes will have a profoundly antithetical, long term impact on national engineering work force equity.

Retention in Engineering

The inveterate retention problem of minority students in engineering colleges is an embarrassment to all of us working in higher education. The attrition of minorities has hovered at about twice the attrition of non-minorities for almost two decades. Two-thirds of all minority students who enroll in engineering do not get a degree in engineering. And we seem to be able to solve the problem, but only on a small scale. There are a number of very successful, individual engineering programs. For example, NACME's programs have a retention rate greater than 80 percent. Our relatively new Engineering Vanguard Program, which targets inner city youth whose circumstances led them to be underachievers in high school and which has a broader goal of expanding the pipeline for science-based disciplines, has not lost a single student in five years. But all of NACME's undergraduate programs collectively reach only 1,200 minority students each year, and there are more than forty thousand enrolled in engineering colleges. All of the successful programs together have not made a dent at the national level.

Perhaps, what's most frustrating is that those of us in the business believe we know what impediments exist, as well as what the solutions are. At the top of the list of barriers is unmet financial need. Many minority students in engineering college are operating at the edge of financial viability from year to year and are continuously at risk of dropping out. If anything unexpected happens, e.g., if tuition rises substantially or their financial package is decreased — all too frequent occurrences — they're forced to leave. Exacerbating this problem is shifting financial aid policy at universities. In particular, scholarships are increasingly used as a recruitment tool to maximize net tuition revenues.

Reinforcing a problem described earlier, some universities widely use SAT scores to determine "merit" for purposes of scholarship allocations. Since scores are highly correlated with family income, this, whether a conscious policy or not, supports the paradoxical university strategy of attracting wealthy students through scholarships. As in retail operations, the effective use of discounting leads ultimately to increased market share and higher revenues. In this case, students from high-income families constitute the desired market.

The financial needs of poor students are increasingly met through loans rather than scholarships, and the debt burden for minority students has been growing at more than twice the rate for non-minority students. This has a major impact on retention. A study by the General Accounting Office shows that a shift of just $1,000 from scholarship aid to a loan for students from families in the lowest income quartile reduces their probability of graduating by 17 percent.

Other issues affecting retention include institutional factors, the social construction of the discipline — a white male dominated culture, low expectations and unsupportive attitudes of engineering faculty towards minority students, gender and ethnic isolation in engineering college, the lack of mentors and the absence of peer support. These are not necessarily the result of overt or even conscious efforts to discriminate. They simply reflect our ethnocentric culture and lack of experience, sensitivity and effectiveness in cross-cultural interactions. These are all things that can be substantially eliminated through professional development and training of faculty and students.

Insufficient academic preparation is also a factor in the attrition equation. Unfortunately, virtually all of the retention efforts focus on this issue to the exclusion of all others. Preparation is certainly important. Minority students from inner city schools, for example, are often less well-prepared than their peers from suburban schools or from the specialized mathematics and science magnet schools, which are increasingly popular among students interested in engineering. Though academic support for some students is essential, more resources need to be devoted to the other factors that play an important role in the overall attrition rates. Moreover, a concentration on presumed student deficiencies can be highly counterproductive. It reinforces low expectations, skepticism and negative attitudes towards the students, and neglects the fact that many minority students who leave engineering are doing quite well academically.

Pre-college Preparation

The linchpin of the crisis we've been describing is that public schools fail to prepare minority students for higher education and particularly for math and science-based majors. The demographics of elementary and secondary school make this issue extremely compelling. Underrepresented minority children constitute more than 30 percent of total enrollment of all American schools, and 54 percent in Central Cities, as defined by the Census Bureau. Despite billions of dollars invested annually in education reform, based on the National Assessment of Education Progress, very little progress has been achieved for students from all ethnic groups. And though the gaps between minorities and non-minorities in mathematics and science achievement have narrowed slightly, they still loom very large.

At the root of the problem is an enduring belief in widely varying innate and immutable intelligence among children. Belying cognitive research outcomes, the American education establishment remains unwilling to internalize the idea that all children can learn at high levels, that academic achievement is more a function of effective effort than genetics. The pervasiveness of tracking systems continues to disproportionately assign minority students to lower level classes. Moreover, low expectations of students not in the highest tracks deny opportunity to the vast majority of non-minority students as well. No amount of spending, curriculum reform, national testing or learning technology will solve our education problem as long as teachers believe that students can't learn. A few years ago, a fraction of the population with a high quality mathematics and science education sufficed to meet the work force needs, but in today's world that will not serve either the work force or the national interest well. Still, only 15 percent of American students take the academic mathematics and science curriculum required to major in a technical field. For minority students, it's six percent. In contrast, calculus and physics, which are not even available at many American high schools, are required courses in many of our competitor nations. No less than a fundamental paradigm shift away from the innate model of intelligence can successfully address this problem.

Conclusion

We cannot successfully continue down the path we're headed as a nation: bifurcated into haves and have-nots, stratified by ethnicity, with limited access to upward mobility through higher education. We cannot afford the huge gaps in access to engineering careers, traditionally the most effective path to upward mobility, with 28.5 percent of the college-age population generating fewer than ten percent of the bachelor's degrees and fewer than three percent of the doctorates.

Though these outcomes by themselves can be discouraging, we shouldn't lose sight of the fact that our gains have been impressive. We must not lose sight of the momentum we've achieved and the progress we've made over the past quarter century — from producing just a few hundred minority graduates per year at the end of the Civil Rights era, we're now producing more than 6,400; from a handful of doctorates annually, we now award a couple of hundred. But, these gains are quite tenuous and the infrastructure that supports them is very fragile. Given that few research and development operations and few universities have a critical mass of minority engineers, eliminating policies that create equitable opportunity and inclusion can wipe them out overnight. This vulnerability is what makes the risks associated with the effectively orchestrated anti-affirmative action movement so alarming.

America's leadership in all sectors has a responsibility to take back the ownership of civil rights policy and its legitimate goals — neither preferential treatment, nor pretense of a color-blindness — but genuine equal opportunity through proactive, *affirmative*, action. We must reaffirm our commitment to equity. We must raise our voices to a level that can compete with the increasingly strident voices of opposition. Today, the American people are hearing loudly and clearly only one side of the story. I'm convinced that, once we articulate the other side, once we present our case with the same degree of intensity, coherence and effectiveness, once the American people understand the real consequences of backsliding on matters of social justice — the impact on our economic development, the effect on our standard of living, the degradation of our political stability, and, ultimately, the threat to our freedom — they will collectively make the correct policy decisions and the right ethical and moral choices.

Data Sources

1. *Chronicle of Higher Education,*
    http://www.chronicle.com/
2. Commission on Professionals in Science and Technology, *Professional Women and Minorities: A Total Human Resources Compendium* Twelfth Edition, E. L. Babco, Washington, DC, 1997.
3. Council on Competitiveness, *Challenges*, Washington, DC, June/July 1997.
4. Engineering Workforce Commission, time series databases on engineering enrollments and graduations collected under a grant from NACME, Inc.
5. NACME, Inc., *NACME Research Letter*, "Patterns in the Production of Minority Engineers," P. Marcotullio, New York, NY, May 1997.
6. NACME, Inc., *NACME Research Letter*, "Bridging the Ethnic and Gender Gaps in Engineering," G. Campbell Jr., New York, NY, May 1996.
7. NACME, Inc., *NACME Research Letter*, "Retention of Minorities in Engineering: Institutional Variability and Success," C. Morrison and P. Marcotullio, New York, NY, December 1995.
8. NACME, Inc., *NACME Research Letter*, "Uninformed Decisions: A Survey of Children and Parents about Math and Science," R. Leitman, K. Binns and A. Unni, New York, NY, June 1995.
9. NACME, Inc., *NACME Research Letter*, "Minority Graduation Rates, Comparative Performance of American Engineering Schools," G. Campbell Jr., R. Denes, D. L. Friedman, L. Miyazaki, New York, NY, December 1991.
10. NACME, Inc., NACME Conference on Research and Pol-

icy: Minorities in Science, Engineering and Mathematics, "Financing Opportunity for Postsecondary Education," T. Mortenson, New York, NY, 1995.

11.   National Science Board, *Science and Engineering Indicators - 1996*, Washington, DC, NSB96-21, http://www.nsf.gov/srs/sbe/seind96/start.htm

12.   National Science Foundation, 1996, *Women, Minorities, and Persons with Disabilities in Science and Engineering: 1996*, NSF 96-311, Arlington, VA.

13.   *The Official Guide to Racial and Ethnic Diversity,* C. Russell, New Strategist Publications, Inc., New York, NY, 1996.

14.   U.S. Department of Commerce, Bureau of Economic Analysis, http://www.bea.doc.gov/

15.   U.S. Department of Commerce, Bureau of the Census, http://www.census.gov/

16.   U.S. Department of Education, National Center for Education Statistics, http://www.ed.gov/NCES.

17.   U.S. Department of Labor, Bureau of Labor Statistics, http://www.bls.gov/

George Campbell, Jr. is Former President of Cooper Union, Former President and CEO of NACME, Inc., and Former member of the President's Information Technology Advisory Committee, Socioeconomic and Workforce Panel.

RACE and FEAR
The Real Hot Buttons Behind the Diversity Debate

by
The Late William H. Gray, III

On the major challenges to diversity...

A major challenge is fear - fear that with change, there are those who are going to lose - lose jobs, lose contracts, lose status, lose seats in higher education and graduate schools, lose scholarships. Those who exploit that fear, for whatever purposes, present the greatest challenge to diversity. Another challenge to diversity is the lack of resources to achieve it. There are groups of citizens - especially in higher education - who are highly qualified to excel but who don't have the resources to participate. Thus, the ability to diversify often is limited by financial constraints.

On the role of economics...

I think economics plays a part in fear, especially during times of uncertainty. Even though there's been a pretty strong economy for the last year or so, its clear from all data that most Americans still feel pretty uncertain about their economic future. And that uncertainty is brought on by a rash of monumental changes taking place - changes in the economy, changes in the world economic structure, changes in geopolitical structures that have an impact on Americans.

But, really, the fundamental question that we are facing in diversity issues in this country goes beyond economic uncertainty; it goes to the bedrock of who we are. Probably the most emotional, gut-wrenching issue of American history is race.

America has gone through a variety of economic changes; we've survived depression, recession, ups, and downs. But race is the one issue that we have never fundamentally been able to deal with. It's the most volatile and most painful issue in our history. We fought America's greatest war not in Vietnam, not in Korea, not in World War II or World War I, not even in the American Revolution. It was the Civil War, and it was fought over the issue of race. Yes, there were economic, political, and cultural consequences, but they weren't the causes; they were the effects. The real cause of that Great War was race. And then we ended up having American apartheid for another 100 years anyway. Today, we're only 30 years away from the climate of legally sanctioned racism.

Here's the problem: Diversity brings into focus all of that painful, 300-year history and all of its implications. The reason it does is because the diversity issues of the late 20th century all involve race. One hundred years ago, diversity was not race-based, it was ethnic - it was immigrants for Europe, it was the Irish, the Italians, the Jews. But now we are facing a diversity that is characterized to a great extent by significant differences of both color and culture; it is Afro-centered, Hispanic, Asian, and Native American. So when we talk about diversity in the 20th and 21st centuries, who are we talking about? We're talking about people of color. Racism and fear are the major issues in this diversity - even more than economic uncertainty.

On where the debate is taking place...

It's very interesting to note where the debate about diversity is taking place. It is taking place primarily in the political arena. Here at the College Fund, we have a lot of contact with top corporate leaders; none of them is taking about getting rid of those instruments that produce diversity. In fact, they say that if their companies are to compete in the global village and in the global marketplace, diversity is an imperative. They also say that the need for talented, skilled Americans means we have to expand the pool of potential employees. And in looking at where birth rates are growing and at where the population is shifting, corporate America understands that expanding the pool means promoting policies that empower and provide skills to more minorities, more women, and more immigrants. Corporate leaders know that if that doesn't occur in our society, they will not have the engineers, the scientists, the lawyers, the business managers, or the accountants they will need.

Likewise, I don't hear people in the academy saying, "Let's go backward. Let's go back to the good old days, where we had a meritocracy" (which was never true - we never had a meritocracy, although we've come closer to it in the last 30 years). I recently visited a great little college in upstate New York - Geneseo, part of the State University of New York system - where the campus has doubled its minority population in the last six years. I talked with an African American who has been a professor there for a long, long time, and she remembers that when she first joined that community, there were fewer than a handful of minorities on campus. Now, all of us feel the university is better because of the diversity. So where we hear this debate is primarily in the political arena and in the media - not in corporate board rooms or on college campuses.

On the realities underlying the debate...

When I hear the debate, it reminds me of what I learned growing up in the South: that when people prey on fear, there is no substance. According to the Glass Ceiling Commission's recently released study, in the 25 years since the late President Nixon mandated affirmative action, progress has been minimal in the private sector. In the top 1,000 corporations in America, women and minorities account for only 3 percent of the executive leadership. When you stretch it to the top 2,000 corporations, it jumps an amazing 2.5 percent, to 5.5 percent. So who's taking whose jobs?

Where the commission reported more progress was in middle management, and there, the progress was made primarily by white women, who account for 40 percent of the managers. In contrast, black males account for just 4 percent, and others, for less than that. So the fear and debate that center around the argument that someone is taking some other, more qualified person's job because of affirmative action just aren't statistically supported. And neither are the fears and arguments about college admissions. The number of blacks at white colleges in America averages only 6 percent nationwide. If you add Hispanic, Asian Americans, and Native Americans to the mix, minority presence on white campuses jumps up into the mid-teens. The numbers show that no one is taking anyone else's place.

Then there are the arguments about preferences and standards. I recently met the president of an Ivy League institution who told me That the university had done a review of its minority students' SAT scores and had found that the average minority student scored at the 50th percentile of the white students. This flies in the face of those who argue that affirmative action is lowering entrance standards: If those minority students didn't deserve to be at that Ivy League institution, then about 40 percent of the school's white students also didn't deserve to be there.

Thus, the statistics prove that whether in the private sector or in the academy, the competition has increased only slightly. And that increase primarily has come *not* from racial minorities, but from women. The confusion over increased competition from minorities versus in-creased competition from women was made evident by a recent *USA Today* poll, which asked, "Are you in favor of affirmative action for racial minorities?" More than two-thirds of the respondents said "No." But when asked if they were in favor of affirmative action for women, two-thirds of the respondents said "Yes." Therefore, I believe the issue here tends to be one of perception; it's emotional, and it tends to focus on the hot button of American history: race.

On "self-separation" on campus...

The so-called self-separation issue is again one that applies only to racial minorities. No one ever says a word about ten Catholic kids having lunch together. No one ever questions why several Jewish stu-dents sit together at the table in the cafeteria. No one raises a fuss about Hillel House, Newman House, Presbyterian Synod House, or Methodist House. Why is it that if ten black kids want to room together in a dorm, it becomes "self-separation" or "separatism?" Why is it that if a bunch of black kids sit down together at a lunch table, it becomes such a threat that we have to give it a label and question why our black college students are separating themselves? The reason, once again, is race. We create special standards for black kids, and we label them in dehumanizing, patronizing ways. We don't do that to other ethnic groups.

When I was a student at Franklin & Marshall University, there were 12 fraternities. There were five black students on the entire campus, and we could join only one fraternity - a Jewish fraternity that was willing to accept blacks. The rest of the fraternities were all white, but nobody made anything of that.

Now, all of a sudden, in the last few years, people have become terribly concerned and have begun to say how un-American it is for black students to want to room together or to want to sit together. At the University of Pennsylvania, there is a residence hall called the Du Bois House, named after W.E.B. Du Bois (who, by the way, graduate from the University of Pennsylvania but was not allowed to be a professor there). Du Bois House wasn't intended to be black, but black students, attracted by the name, started to bid for that dormitory. Before long, the dorm was 90 percent black, and people started saying, "This is segregation." The black students who lived in the dorm accounted for less than 10 percent of all the black students on campus; the other 90 percent were living in integrated dorms with white students.

My point is simply the focus on perception: Why are people so threatened by Du Bois House? I'm not threatened every time I see a group of whites sitting together with no blacks. Why do we expect kids - white or black - who grow up in a largely segregated America, live in largely segregated neighborhoods, go to largely segregated churches, and belong to largely segregated clubs to suddenly want to live with one another the second they arrive on campus? What is amazing is how *little* conflict we have when these youngsters, the vast majority of whom have lived segregated existence, come together at universities. I find that at most institutions, integration *has* taken place. Are there some students who prefer the company of other African American students because they perceive the university as a hostile environment (and sometimes it is), because words like "nigger" are scrawled in the bathrooms and terrible things are written in the newspapers about them? Yes. But the fact of the matter is that they're very small in number. Why do we apply a double standard? Why don't we say anything about white kids who cling to ethnicity or religious heritage to define who they are on campus? We don't. So why do so for blacks?

I do have a problem with the language, a lot of which is cryptic jargon used by those pro and con. For example, we call the debate over affirmative action a "liberal" versus "conservative" debate. Richard Nixon was not, I think, a liberal. But he was the one who instituted specific goals and timetables back in 1970. That's where affirmative action originated.

The issue of "race-based scholarships" also seems to be about language, but really, it's about race. In America today, race-based scholarships, where race is the sole determining factor, account for less than 1 percent of all graduate and undergraduate scholarships, and about 5 percent of all scholarships exclusively at the undergraduate level. Scholarships that include race as a component account for less than 5 percent of all scholarships in America. Yet 9 percent of all scholarships in America are based on religion. There are scholarships for descendants of Confederate soldiers. I don't think too many blacks would be eligible. There are Daughters of Norway scholarships. I don't think too many Hispanic women would be able to apply. Why is it that we worry only about race-based scholarships?

And what is political correctness? What is political incorrectness? The words have no meaning to me. *People* define who they are. We once thought of America as a "melting pot." Later, we thought about America as a "pluralistic society." Now people talk about "multiculturalism." And those people who don't want to face up to the demographic changes in America are talking about going back to the "values of the past."

But the values of the past kept me out of Duke and Arkansas and 80 percent of America's other universities. So when I hear talk show hosts and politicians urging us to "get back to good old American values," I ask "What values are you talking about?" Segregation and sexism are not good values. Do you want to go back to the values Elizabeth Dole talked about in her autobiography, when the first day she went to work at a law firm, a male turned to her and said, "You're taking up the space of a qualified man"? Are those the values we're talking about? I don't want those values. I don't want the old American under any circumstances.

The old America was not the best America. This is a better America, where we have more opportunity for more people, and where our institutions and our marketplaces are reflecting a little bit more - a very little bit more - of what America is really about. I heard on famous reporter who happened to have a severe physical disability talking about going back to an America that values merit, and I thought to myself, "Do you realize what you're saying? Twenty-five years ago, society would have barred you from becoming a syndicated columnist because of your disability. The reason you are a syndicated columnist is that we got rid of the false value that said, 'If you are disabled, you can't possibly think, speak, or write,'" Yet here was a person leading the charge in the name of conservative politics and a conservative agenda that attacks diversity and calling multiculturalism a politically correct phenomenon that has killed American values and merits. What people really are fighting about is, at its core, a significant change in American society that involves the issue of race.

Why are we having this argument? The real issue ought not to be about language, but about our future: If America is going to be greater in the 21st century than it has been the 20th century, *all* of us have to develop skills to be included in this society. It's as simple as that.

On what higher education leaders can do...

I think they've got to help the people of the academy understand the need for inclusiveness, and that being inclusive does not mean lowering standards. Instead, it means looking for qualified people who can contribute to the community and to society, and it means recognizing that *how* we judge those people is an ever-evolving process. Higher education leaders must constantly assess their financial assistance programs to make sure they are providing financial aid to those who most need it. And they've got to explain how those aid systems work. They've also go to explain to the student body and to the faculty the institution's mission and how it works - that even though the campus may be located in some isolated spot, it's going to reflect the population that makes up the nation. And that's essentially what they have to do.

In summing up...

The real question is: In 30 years, have we made so much progress that we no longer need special efforts to redress past and present discrimination based on race and gender? I would say, clearly and unequivocally, "No." And I think most Americans, when pushed, will admit that. The question then becomes: What are the appropriate remedies? You can't apply an economic class remedy to what is a race problem (even though it is reflected in class). That's like treating people who have cancer for pneumonia. Now, they may have pneumonia as a side effect of the cancer, but if you treat only the pneumonia, and not the cancer, they will not be cured.

And that's one of the confusions behind the current debate. Many people say, "Look, we've made enough progress in the past 30 years that really, the problems, the behaviors, and the dysfunctionality that we see are class oriented. They're class oriented, not race-oriented, not gender-oriented." But they fail to reveal the cancer, and they fail to acknowledge that the remedy to cure the cancer is a medicine called affirmative action. And, further, they fail to tell people the truth about the medicine: that historically, statistically, and any other way you want to look at it, the medicine does *not* create reverse discrimination.

The real bottom line, though, is that we ought to be looking at this issue from the other side. Diversity should not be seen as a problem with which we have to deal. Diversity is really our greatest opportunity. It's how we make America stronger, not weaker. It only takes a quick look at the world - through the prism of reality - to see that.

In the next ten years, more than half of all new entrants into the American work force will be minorities. Another 35 percent will be women and new immigrants. If we don't educate *all* Americans to world class standards, we simply will not be able to compete in the global market. Corporate America has recognized this essential fact. As recently as May 1995, Procter & Gamble Chairman and CEO Ed Artzt said in a major speech, "Diversity is an integral part of the character of our company. It is very real to us. It gives us unity. It gives us strength. And it gives us talent - the richness of talent that we need to successfully sell our products to people of all cultures in every market of the world.

We all need to learn this lesson.

*The Late William H. Gray, III was Former Majority Whip of the US House of Representatives and Former President of the United Negro College Fund. The above are excerpts from an interview he afforded to the **Educational Record's** editor-in-chief.*

Are We Still A Land of Opportunity?

by
The Late Charles M. Vest

When we think of the future, scientific and technological innovations often come to mind. But the quality of our future will have even more to do with human relations than it does with science and technology. If this nation is to thrive - economically, socially, politically - we must do all we can to ensure that all of our citizens are able to reach their full potential. Only then will we realize the full benefits to be found in a society peopled with different cultures, races, and nationalities.

Race and Society - One Nation or Many?

In the 1950s and 1960s, we as a nation determined that we would build a racially integrated, nondiscriminatory society, and we recognized that various interim commitments and corrective actions would be required until we reached that goal. Full attainment of that goal has proved more elusive than most anticipated. We now seem to be backing off in many ways - too ideological; too uncomfortable; too difficult.

Educational institutions have had central roles in both the action and debate throughout this period. Fundamentally, this is because of our special responsibility to prepare young people to take their full place in our society. Indeed, America's course in these matters was largely set by the 1954 Supreme Court decision in Brown v. Board of Education that laid the foundation for the affirmative action initiatives of the 1960s by ordering racial integration of public schools with all deliberate speed.

Today, more than 40 years after Brown v. Board of Education, we still find ourselves at the center of discussion, evaluation, and legal decisions about race and diversity. Largely because of explicit actions to increase access to our colleges and universities, most have become much more diverse racially, culturally, and economically. The presence and role of women on our campuses have improved dramatically. Still, most campuses cannot be judged to be broadly representative of the makeup of contemporary America. Statistics regarding most measures of academic success and access of young people to career, professional, and leadership tracks tell us that the goals set in the 1950s and 1960s have not yet been achieved. My sense is that we are losing will, ignoring realities, falling into political partisanship, and, not infrequently, introducing mean-spiritedness into the national debate on these matters.

Effectively addressing issues of race and diversity is too essential to the future of the United States to allow it to be dissipated in partisan rhetoric. Maintaining our momentum is too urgent to allow it to be defined away through narrow, technical judicial decisions. Reinvigorating a national commitment is too demanding to allow it to drown in a sea of red tape. We need both idealism and pragmatism, but we cannot, through what Father Theodore Hesburgh refers to as "combat fatigue," enter the next century without making real progress toward broad equality.

It astounds me how frequently the issue of diversity is addressed as if it were an abstract concept. Racial diversity is a reality of American life in 1996, and we know with certainty that it will be an even more dominant reality in, say, 2015, when the children being born this year are of college age. In 2015, the college-age population of the US will be 16-percent African-American and 19-percent Hispanic-American, and the mix of new immigrants to our shores, especially from Asia and Southeast Asia, also will contribute more substantially to the makeup of our citizenry.

By the year 2015, the work force will be one-third white male, one-third white female, and one-third people of color. All these workers will be toiling to support not only themselves, but all of us who, as retirees, will be dependent upon them - and they will constitute a much smaller proportion of our population. (In 2015 there will be only half as many people working and supporting the retired population, as there were in 1960). If they do not form a cohesive, productive society, the future will indeed be bleak. This prognostication is truly daunting, especially when combined with the fact that we will need to compete in a marketplace and economy that will be even more globalized and integrated than today.

Thus, even if we are willing to ignore the historical imperative and noble goal of equality and true integration, we must be problem solvers and set a sound course for our rapidly changing nation.

It is sorely tempting to declare victory and turn our back on affirmative action and related processes in America. How pleased I would be if we could legitimately assume that all of our citizens have reached a sufficient state of actual equality of opportunity and access that we could adopt simple, race-blind approaches to all that we do. That, of course, is the goal. But is it an honest evaluation of the situation today?

One need only peruse the extensive tabulations of national statistics regarding wages, crime, education, health, and many other parameters in Andrew Hacker's book, *Two Nations: Black and White, Separate, Hostile, Unequal*, to know that we have not achieved anything approaching equality across the racial boundaries of our society. If that is not convincing, read the front page of any urban newspaper on any given day.

Yet we are retreating. The federal district court ruling in Hopwood v. University of Texas has had repercussions around the country - as organized efforts to end affirmative action continue to grow. The actions of the University of California's Board of Regents are well known; and there have been legislative moves in other states to outlaw affirmative action, or to amend their constitutions.

In this context, I use the term "affirmative action" rather broadly to refer to programs or actions that specifically foster access or participation of minority groups or women in educational programs or jobs. This breadth seems appropriate in discussing universities in light of recent court decisions. MIT's admissions process is consistent with the Supreme Court's 1978 Bakke decision that universities may consider race "as one factor among many" in making admissions decisions. We build our admitted class to bring together students from diverse geographic, economic, cultural, racial, and experiential backgrounds, all of whom have exhibited the intellectual capacity, achievement, and motivation that are needed to succeed and benefit from MIT. Furthermore, our undergraduate financial aid is awarded solely on the basis of demonstrated financial need.

Yet in 1996, in Hopwood v. University of Texas, the Fifth Circuit Court of Appeals effectively reversed the Bakke decision for public institutions in Texas, Mississippi, and Louisiana, by declaring that "any consideration of race or ethnicity by the law school for the purpose of achieving a diverse student body is not a compelling interest" and therefore is not permitted.

I do not wish to defend across the board all federal affirmative action laws and set-aside policies, with their attendant red tape, cumbersome bureaucracies, and often artificial metrics. But I do want to defend the core concept that determined, often race-specific consideration and effort are still essential to move us toward the integrated, cohesive society we will need in the years ahead. The society I believe we will need is one in which individuals can realize their potential, and in which we can draw effectively on the individual and collective strengths and talents of our citizens of all colors and ethnicities. We cannot command, decree, or wish into existence such a nation. Rather we must work proactively to build it through the environments and opportunities we create for learning and working.

The idea that affirmative action programs are unnecessary or even unconstitutional is gaining momentum just at a time when we in science, engineering, and higher education are beginning to see some real results from these programs.

In 1996, the American Council on Education released its study on minorities in higher education and reported a record number of Ph.D.'s awarded to black graduate students in 1995. And over the past eight years, the National Science Foundation reports there has been a 75-percent increase in the number of science and engineering doctorates awarded to black graduate students - from 319 in 1987 to 557 in 1995. The media and others have hailed this as a dramatic increase. It is, indeed, real progress; nonetheless, the absolute numbers are stunningly small. Last year, for example, the number of blacks receiving the doctorate in electrical engineering in the US rose 40 percent over the previous year - to 24. Yet this is out of a total of 966 doctorates awarded in that field.

And yet there are arguments over the reasons for this progress. Supporters of affirmative action claim the increase as evidence of the programs' effectiveness, while critics argue that it is the result of increased educational opportunities, and that any benefits of affirmative action are offset by the negative effects of what they regard as preferential treatment of minorities.

Did "affirmative action" play a role in this modest success? It should not be a difficult matter to assess how many of these new Ph.D. graduates were definitively encouraged or enabled to reach this high level of attainment by specific programs or support. It should not be a matter of guesswork; the data should be obtained and affirmative action and outreach programs should be objectively evaluated on the basis of outcomes over time. It should not be a matter of ideology of the left or of the right. We should assess where we are, demonstrate what does and doesn't work, and get on with the job.

In the current legal environment, attorneys are recommending to organizations that were established specifically to promote educational opportunity for minority students that they modify their eligibility criteria to indicate that they will review applications without regard to the applicant's ethnicity. Frankly, this strikes me as a strange and artificial approach.

My own view is that we must hold to our principles if our nation is to benefit from the full range of talent needed to meet the challenges of a changing world. Our journey is not over. Our goal is not attained.

I believe that the time will come when affirmative action programs will no longer be necessary, but for now, we still have a compelling need for proactive efforts, despite calls by some that what is needed instead is simply stronger enforcement of antidiscrimination laws. Indeed, as Tom Wicker put it in his recent book, Tragic Failure: "If enforcement of antidiscrimination laws is the alternative to affirmative action, race, sex, and ethnic discrimination will be with us for a long time."

An Open Society - To Whom?

Race is not the only focus of the argument about how open our society should be. These are economically difficult times in America - at least relative to our aspirations and to the post-war boom years. And as times get tight, there is a natural tendency to turn inward. So once again, we hear concerns that we should not be educating so many foreign graduate students. We hear that immigrants are a major cause of our woes. And we keep pulling apart into homogeneous groupings of one sort or another. But just because these are natural or understandable tendencies does not make them right.

America has always been a nation of immigrants and we have always been a land of opportunity. These statements perhaps sound quaint or old fashioned, but they are true, and we must retain their spirit.

Each year, my wife Becky and I host a dinner in our home for the men and women who are retiring from the tenured faculty ranks of MIT. These are always extraordinary assemblages of talented and accomplished colleagues - people who have defined MIT, and who have defined their professional and scholarly fields. No lack of bold thought there!

Yet, as I survey that room each spring, I realize how much MIT and indeed America have benefited from our being open to those from other countries, and how wise has been our tradition of selecting and advancing people on the basis of their talent and accomplishment rather than their wealth or nationality.

Now, some might say that this represents a passing era, that what I am observing has its origins in the intellectual migrations from Europe associated with the turmoil of the World War II era. Or it might even represent the vestiges of the times during which the leading universities in science and engineering were in Germany and England.

No, it is an ongoing fact that the excellence of our institutions is due in very large measure to our openness to international scholars. MIT faculty who have received the Nobel Prize include individuals who were born in Japan, India, Italy, and Mexico. Our provost was born in Israel. We have deans who were born in Canada and Australia. Almost all came to the US as graduate students.

In fact, about one-third of all Ph.D. degrees in science and engineering earned in US universities are awarded to foreign citizens. (In engineering alone, half of the Ph.D.'s are earned by foreign citizens.) Many of these doctoral recipients initially pursue their careers in the US, and about 40 percent of them appear to remain here permanently. What a magnificent resource for our industries, universities, and government laboratories. Openness and meritocracy are what have made our universities great, and we must continue that spirit and philosophy in our national endeavors.

At the same time, we should concentrate on improving both science education and general education in this country's K-12 system in order to increase the number of motivated, well-prepared students entering universities and colleges. We should value more highly intellectual pursuits and celebrate the accomplishments of those who contribute to our health and quality of life by advancing science and technology. This is the way to ensure that, in the long run, our graduate programs have a larger, more stable base of US students.

We must, however, continue to provide access, opportunity, and welcome to the brilliant immigrants who contribute so much to our society - people like Institute Professor Hermann Haus, who received the National Medal of Science in 1996. Recollecting the call from John Gibbons, the President's Science Advisor, Professor Haus said, "I did not trust my senses, at first.

After the news sunk in, the thoughts that came to my mind were that I was grateful to my fate for having come to the US, a victim of the 1945 ethnic cleansing in Yugoslavia; for becoming a citizen; and for the recognition I received on account of work I thoroughly enjoyed and for the privilege of association with outstanding students and colleagues."

I can think of no more eloquent description of what it means for this country to be the land of opportunity. We must retain our commitment to this bold dream.

The Late Charles M. Vest was MIT President, and a member of the Clinton Administration's President's Committee of Advisors on Science and Technology.

THE WHITE HOUSE

WASHINGTON

October 23, 1998

Fact Sheet: National Science and Technology Council (NSTC)

Background. President Clinton established the National Science and Technology Council (NSTC) by Executive Order 12881 in November 1993. The NSTC is a cabinet-level council that coordinates R&D policies and activities across the federal agencies. It consolidates the responsibilities previously carried out by a number of interagency councils, including the Federal Coordinating Council for Science, Engineering, Technology, the National Space Council, and the National Critical Materials Council.

Composition. The president chairs the NSTC. Members include the vice president, assistant to the president for science and technology, cabinet secretaries and agency heads with significant science and technology responsibilites, and senior White House officials.

Responsibilities. The major functions of the NSTC are to:

- Coordinate the formulation of S&T policy
- Ensure S&T policy decisions and programs are consistent with the president's stated goals
- Help implement and integrate the president's S&T policy agenda across the Federal government
- Ensure S&T are considered in the development and implementation of all Federal policies and programs
- Further international cooperation in S&T

Organization. NSTC accomplishes much of its work through the following five committees:

- Committee on Environment and Natural Resources
- Committee on International Science, Engineering, and Technology
- Committee on National Security
- Committee on Science
- Committee on Technology

Each committee is further organized into subcommittees and interagency working groups. The NSTC also creates cross-committee working groups as needed to review and coordinates specific policies or programs that span the interests of the standing committees. The cross-committee working groups currently include those on Federal Laboratory Reform, Health Preparedness for Future Troop Deployment, Global Positioning System, and Aviation Safety and Security.

President Clinton Decision Directives. President Clinton issued Presidential Decision Directives (PDDs) to the NSTC requesting recommendations on a variety of topics, including national space transportation, emerging infectious diseases, achieving greater diversity in the scientific and technology work force, and federal laboratory reform. The NSTC web site provides a complete list of PDDs and the resulting NSTC reports during the Clinton Administration.

Clinton Administration Reports. Coordination of multi-agency reports is an important role of the NSTC. Since January 1997 NSTC has published the following:

- *National Science and Technology Council 1997 Annual Report* (April 1998)
- *Our Changing Planet: The FY 1999 U.S. Global Change Research Program* (March 1998)

Edited and Foreword by Oliver McGee, Ph.D.

- *Program Guide to Federally Funded Environment and Natural Resources R&D* (February 1998)
- *National Plant Genome Initiative* (January 1998)
- *Technologies for the 21st Century, Supplement to the President's Budget* (November 1997)
- *Transportation Science and Technology Strategy* (September 1997)
- *Interagency Assessment of Oxygenated Fuels* (June 1997)
- *Our Changing Planet: The FY 1998 U.S. Global Change Research Program* (June 1997)
- *Science and Technology: Shaping the Twenty-First Century* (April 1997)
- *Investing in Our Future* (April 1997)
- *Manufacturing Infrastructure: Enabling the Nation's Manufacturing Capacity* (April 1997)
- *Computing, Information and Communications Technologies for the 21st Century* (March 1997)
- *Status of Federal Laboratory Reforms* (March 1997)
- *Integrating the Nation's Environmental Monitoring and Research Networks and Programs: A Proposed Framework* (March 1997)

Members and Protocol:

The President
The Vice President
Secretary of State
Secretary of the Treasury
Secretary of Defense
Secretary of the Interior
Secretary of Agriculture
Secretary of Commerce
Secretary of Labor
Secretary of Health and Human Services
Secretary of Transportation
Secretary of Energy

Secretary of Education
Secretary of Veterans Affairs
Administrator, Environmental Protection Agency
Director, Office of Management and Budget
Chair, Council Intelligence Agency
Assistant to the President for National Security Affairs
Assistant to the President for Science and Technology
Assistant to the President of Domestic Policy
Assistant to the President of Economic Policy
Director, Arms Control and Disarmament Agency
Administrator, National Aeronautics and Space Administration
Director, National Science Foundation
Director, National Institutes of Health

For more information see:
http://www.whitehouse.gov/WH/EOP/OSTP/NSTC/html/NSTC_Home.html
Or call the Office of the Executive Secretary at (202) 456-6104.

Edited and Foreword by Oliver McGee, Ph.D.

THE WHITE HOUSE

WASHINGTON

October 23, 1998

Fact Sheet: President's Committee of Advisors on Science and Technology (PCAST)

Background. President Clinton established the President's Committee of Advisors on Science and Technology (PCAST) by Executive Order 12882 in November 1993. The committee advises the president on the administration's science and technology budgets and policies. PCAST meets in public session an average of four times a year.

Composition. PCAST consists of 19 members and includes distinguished individuals from industry, education, research institutions, and other non-governmental organizations. The president appoints all members. The assistant to the president for science and technology co-chairs the committee with a private sector member selected by the president.

Responsibilites. The responsibilities of PCAST are "to advise the president on issues involving science and technology and their roles in achieving national goals, and to assist the National Science and Technology Council (NSTC) in securing private sector participation in its activites."

Organization. PCAST members accomplish some of their work in ad hoc groups. Currently, PCAST working groups are addressing the following five topics:

- Security
- Education
- Environment and Natural Resources
- Government Performance and Results Act (GPRA) or OSTP
- International Cooperation in Energy R&D

Clinton Administration Reports. PCAST has published the following Clinton Administration reports:

- *Teaming With Life: Investing in Science to Understand and Use America's Living Capital* (June 1998)
- *Report to the President on the Use of Technology to Strengthen K-12 Education in the United States* (March 1997)
- *Report to the President on Federal Energy Research and Development for the Challenges of the Twenty-First Century* (November 1997)
- *Report on Sustainable Development* (January 1997)
- *Principles on the U.S. Government's Investment Role in Technology* (June 1996)
- *Report on Research Universities* (June 1996)
- *Report on Preventing Deadly Conflict* (November 1996)
- *Science and Technology Initiatives* (December 1996)
- *Report of the PCAST Panel on U.S. – Russian Cooperation to Protect, Control, and Account for Weapons-Useable Nuclear Materials* (May 1995)
- *The U.S. Program of Fusion Energy Research and Development* (July 1995)
- *Science and Technology Principles* (September 1995)
- *Report to the President on Academic Health Centers* (November 1995)

Edited and Foreword by Oliver McGee, Ph.D.

Members:

- **Neal F. Lane** – Assistant to the President for Science and Technology and Director, Office of Science and Technology Policy (*co-chair*)
- **John A. Young** – Former President and CEO, Hewlett-Packard Company (*co-chair*)
- **Norman R. Augustine** – Chairman and CEO, Lockheed Martin Corporation
- **Francisco J. Ayala** – Donald Bren Professor of Biological Sciences, Professor of Philosophy, University of California-Irvine
- **John M. Deutch** – Institute Professor, Department of Chemistry, Massachusetts Institute of Technology
- **Murray Gell-Mann** – Professor, Santa Fe Institute; R.A. Millikan Professor Emeritus
- **David A. Hamburg** – President Emeritus, Carnegie Foundation of New York
- **John P. Holdren** – Teresa and John Heinz Professor of Environment Policy, John F. Kennedy School of Government, Harvard University
- **Diana MacArthur** – Chair and CEO, Dynamic Corporation
- **Shirley M. Malcolm** – Head, Directorate for Education and Human Resources Programs, American Association for the Advancement of Science
- **Mario J. Molina** – Institute Professor, Department of Earth, Atmospheric and Planetary Sciences, Massachusetts Institute of Technology
- **Peter H. Raven** – Director, Missouri Botanical Garden; Engelmann Professor of Botany, Washington University in St. Louis
- **Sally K. Ride** – Professor of Physics, University of California – San Diego

Meeting America's Needs for the Scientific and Technological Challenges of the Twenty-First Century – A Retrospective

- **Judith Rodin** – President, University of Pennsylvania
- **Charles A. Sanders** – Former Chairman, Glaxo-Wellcome Inc.
- **David E. Shaw** – Chairman, D.E. Shaw and Company, and Juno Online Service
- **Charles M. Vest** – President, Massachusetts Institute of Technology
- **Virginia V. Weldon** – Director, Center for the Study of American Business, Washington University in St. Louis
- **Lilian Shiao-Yen Wu** – Member, Research Staff, Thomas J. Watson Research Center, IBM

For more information see:
http://www.whitehouse.gov/WH/EOP/OSTP/NSTC/PCAST/pcast.html
Or call the Office of the Executive Secretary at (202) 456-6100.

Edited and Foreword by Oliver McGee, Ph.D.

About the Editor

Oliver G. McGee III, PhD, MBA, CFRM

Oliver G. McGee III is a teacher, a researcher, an administrator, and an advisor to government, corporations and philanthropy. He is professor and former chair of the Department of Mechanical Engineering at Texas Tech University. He was formerly professor of mechanical engineering and former Vice President for Research and Compliance at Howard University, serving as the chief research officer of the 140-year institution, and as a Cabinet-level executive reporting to the Howard University President.

Primary responsibilities included shaping vision, overall management, and fiscal oversight of Howard University's research strategy, planning, coordination, compliance, administration, and sponsored programmatic operations -- including research policy and administration strategy; human resource staffing; research financial planning, investments, valuation and risk management; sponsored programmatic operations; research marketing, development and communications; and supervision of research capacity growth and cost recovery; and technical assistance programs. As Vice President for Research and Compliance, Dr. McGee provided executive leadership and administrative oversight of Howard University's Office of the Vice President for Research and Compliance, Office of Research Compliance, Office

of Sponsored Programs, and Office of Technology Transfer. He had
fiscal responsibility for central administration oversight of annual
university sponsored grants and contracts expenditures of over $70
million in FY07.

Dr. McGee is former Senior Vice President for Academic Affairs of the United Negro College Fund (UNCF), Inc. The UNCF is the
nation's largest, oldest, most successful and comprehensive minority
higher education cooperative financial security and assistance philanthropic organization. McGee served as the chief academic officer of
the 60-year old College Fund with primary responsibilities of management, oversight, and executive leadership of all UNCF programmatic departments to support the Fund's 39 private Historically Black
Colleges and Universities (HBCU) member institutions. He provided
leadership and administrative oversight of the UNCF Office of Academic Affairs, including several UNCF departments: Scholarships and
Grants Administration, Corporate Scholars, Fiscal and Strategic
Technical Assistance and Academic Programs, Science, Technology,
Engineering, and Mathematics (STEM) Education and Pipeline
Development, The Fredrick D. Patterson Research Institute, and The
UNCF Institute for Capacity Building. He also had fiscal oversight
responsibility for managing an annual UNCF programmatic operating
budget of over $40 million.

Dr. McGee was Professor and former Chair (2001-2005) of the
Department of Civil & Environmental Engineering & Geodetic
Science at Ohio State University. He is the first African-American to
hold a professorship and a departmental chair leadership in the centu-
ry-and-a-quarter history of Ohio State University's engineering
college. McGee is the former United States (U.S.) Deputy Assistant
Secretary of Transportation for Technology Policy (1999-2001) at the
U.S. Department of Transportation (DOT). Appointed by former U.S.
President William Jefferson Clinton, McGee served as the lead direct
report to former U.S. Transportation Secretary Rodney E. Slater, with
primary responsibility for management, oversight, and executive
coordination of technology policy and programs across the ten modal
transportation administrations of the U.S. DOT – totaling approximately $5 billion annually for research and development of the na-
tion's complex transportation systems. McGee came to the U.S. DOT

after serving as Senior Policy Analyst (1997-1999) in the White House Office of Science and Technology Policy (OSTP). He led OSTP's contribution to the President's Initiative on Race, which resulted in the White House policy document, *Meeting America's Needs for the Scientific and Technological Challenges of the Twenty-First Century – A White House Roundtable Dialogue for President Clinton's Initiative on Race.*

At the U.S. DOT, McGee led the interagency team primarily responsible for the development, preparation and coordination of the *National Science and Technology Council's (NSTC) National R&D Plan for Aviation Safety, Security, Efficiency & Environmental Compatibility* – the FAA, NASA & DOD joint plan to implement the $1.3B FY01 R&D investment recommendations of former President Clinton's 1997 Commission on Aviation Safety & Security, chaired by former Vice President Albert Gore. Former U.S. Transportation Secretary Rodney E. Slater at the *"Aviation in the 21st Century – Beyond Open Skies Ministerial"* in Chicago, Illinois launched this plan on December 6, 1999. McGee served as co-chair of the NSTC Committee on Technology Wire Systems Safety Interagency Working Group, which resulted in the White House policy document, *Review of Federal Programs for Wire Systems Safety*, aimed to benchmark agency efforts to optimize Federal R&D leading to a national strategy for wire system safety in response to the Gore Commission on Aviation Safety & Security. He also led the teams responsible for the development and preparation of the 2025 national transportation policy reports, *Transportation Decision Making – Policy Architecture for the 21st Century* and *The Changing Face of Transportation* both released in January 2001. His leadership in this effort resulted in his teams of federal government career staff receiving the *2000 U.S. Secretary of Transportation's Partnership for Excellence Award* – the second highest award within the Federal Department. This award recognizes department-wide inter-modal teams/groups that have used partnership models to support one or more DOT strategic plan goals.

Recognitions and Honors: Dr. McGee has also held a number of faculty appointments and research positions at the Massachusetts Institute of Technology, Georgia Institute of Technology, Ohio State University, and the University of Arizona. He came to higher education with industrial-sector experience through engineering positions

held at NASA's John H. Glenn Research Center at Lewis Field and
Boeing (formerly McDonnell Douglas) Helicopter Company. He has
received numerous national and state teaching and engineering awards
including a 1991 National Science Foundation Presidential Young
Investigator Award, a 1993 NASA Faculty Award for Research, the
Council for Advancement and Support of Education (CASE) & the
*Carnegie Foundation for the Advancement of Teaching's 1995 State of
Georgia Professor of the Year, U.S. Black Engineer Magazine's 1996
Black Engineer of the Year Award, Education College-Level, and
Science Spectrum Magazine's "Fifty (50) Most Important Blacks in
Research Science" for 2004.* In 2005, McGee was named to *Science
Spectrum Magazine's "Top Minorities in American Research Science"*
List. He has authored numerous research journal articles on subjects
ranging from interdisciplinary design synthesis and vibration control
of mechanical and structural systems to aeromechanics and control of
dynamic flow instabilities in air-breathing propulsion systems used for
aircraft. McGee has served on the Board of Editors of the international
journal, Computer Modeling in Engineering and Sciences, and he is a
former member of the National Science Foundation Advisory Committee for Engineering.

Educational Data: Born in Cincinnati, Ohio, McGee is a
graduate of Woodward High School in Cincinnati, Ohio, the oldest
public high school west of the Appalachian Mountains. He is a 1981
graduate of Ohio State University in civil engineering. He received a
M.S. degree in civil engineering in 1983, and a Ph.D. degree in
engineering mechanics (with a minor in aerospace engineering) in
1988, both from the University of Arizona, and a M.B.A. degree in
2004 from the University of Chicago, Graduate School of Business
(GSB). McGee earned the Certificate of Professional Development
(C.P.D.) in 2001 from The Wharton School of the University of
Pennsylvania and he is an alumnus of Wharton's Advanced Management Program (AMP36), and the Private Wealth Management (PWM)
Program (2006), sponsored by The Wharton School and the Institute
of Private Investors (IPI). He has studied at the University of Cam-
bridge (England), Stanford University, The Aspen Institute, Harvard's
Law School and John F. Kennedy School of Government, the Council
for Excellence in Government, the American Association of State

Colleges & Universities' (AASCU) Millennium Leadership Institute Fellows Program, an educational management program for prospective chancellors and presidents, and The Directors' Consortium, an executive program on the fundamentals of corporate governance and board service (jointly sponsored by the University of Chicago Graduate School of Business, Stanford Law School, and The Wharton School). He is also a member of the National Association of Corporate Directors (NACD).

Institutional Data on Past Position: Authorized by a charter of the 39th U.S. Congress to which President Andrew Johnson affixed his signature on March 2, 1867, Howard University opened its doors to students two months later. As the nation's largest predominantly black institution of higher learning, Howard University has produced more black professionals than any other institution in the country. More than half of the nation's black physicians, dentists, pharmacists, engineers and architects are Howard graduates.